FORESTRY FLAVOURS OF THE MONTH

The Changing Face of World Forestry

ALASTAIR FRASER

Table of Contents

To Susie, Ian, Neil, Michelle, Helmi, Aisyah and Ariq

For your understanding during my many absences

Figures

Foreword

Forestry Flavors of the Month is a delightful and compelling read on changes in world forestry spanning more than five decades. What made this book stand out for me, and I believe will for development professionals in general, is how firmly Alastair has anchored the issues and challenges of world forestry in global political economy. He does so without compromising the scientific rigor or focus on human welfare, particularly of the disadvantaged forest-dependent rural communities in the developing world.

Using his more than five decade professional journey as the canvas, Alastair has effectively and efficiently painted for us the ups and downs of world forestry. Book traces the historical shifts in world forestry from a production and plantation forestry focus in 1990's to a paradigm shift to forest-ecosystem services approach for addressing global poverty reduction and climate change challenges, particularly since early 2000's. This discourse takes us from forestry research in "wind tunnels" in U.K. to multitude of global and local initiatives for sustainable forest and plantation management in the Amazon, Brazil, Greece, Indonesia, Surinam and lastly to the socio-ecologically rich continental Southeast Asia.

Alastair does not shy away from clearly laying out the failures of international development assistance and donors in contributing to the current dismal condition of world forestry in general and tropical forests in particular. Furthermore, in using a forestry lens to trace and diagnose global developmental challenges Alastair draws our attention to a critical and persistent problem of short-termism that has bedeviled global sustainable forestry and development initiatives and investments

In addition to foresters, the development practitioners will be well served by this book. It lays bare the adverse consequences of the short-termism curse for the global environmental and human welfare. Alastair, challenges development professional community to find ways and means of reconciling the long-term time frames needed for sustaining forests and economic development with the short-term time frames within which political, economic and technical agencies function.

Having had the privilege of working with Alastair Fraser for last decade and a half in Asia and benefitting from his expertise and experience, I believe this book will by equally, if not more, enriching to foresters of today and tomorrow. It should inspire current and future generations of forestry professionals. I hope this book will guide and inform the next generation of international forestry and development professionals in equipping themselves with the knowledge and skills that will be needed if global environmental and developmental challenges are to be met with greater success in the future.

Javed H Mir
Former Director, Southeast Asia Department
Asian Development Bank

Introduction

Dawn somewhere over Bangladesh and we are woken for breakfast by a stewardess with a tray of orange juice. Long haul flights by oneself can be enlivened if one is fortunate enough to have someone in the neighbouring seat who is interested in chatting. Over the years of travelling to many parts of the world, I have found myself sitting next to an Ambassador, various types of business person, an accountant, a geologist, an engineering consultant, a retailer and many others. This list shows that the conversation often turns to "What do you do"? In my case my response is "I am a forester", and the response is equally often "Oh! that sounds interesting - do you plant trees"? I have never met anyone with any real idea of what such a job entails. Monoculture tree plantations and tropical deforestation are widely held perceptions of what forestry is all about. There are a number of reasons why this may be so, and this book sets out to describe the range of activities that I have been engaged in and in the process give people outside the profession an idea of the sort of things that a forester gets up to during a working career spanning about fifty five years.

It is not intended primarily as an autobiography, nor as a scientific treatise, but instead reflects on some of the topics that I, as a forester working internationally, have had to deal with and interesting or important issues raised by many of the assignments. It also gives a view of what the forestry profession as a whole has been doing during the period from 1960 to the present. The title for the book came to me one day, when chatting to a friend about what I had been up to, and it seemed to me that over the years I had been involved in a very wide range of issues, almost all of which had relatively quickly become a 'hot topic' and then almost as quickly had disappeared and been replaced by some new 'hot topic'

As I write this, the subject of global warming and climate change seems to be the 'hot topic' that is involving many foresters, because forests play an important role in the atmospheric exchange of oxygen and carbon dioxide, and the latter is considered one of the main culprits in creating a greenhouse effect that is contributing to atmospheric warming. It is eight years since Al Gore presented his documentary *An uncomfortable truth* and already the world seems to be getting bored of the subject. Recent opinion polls suggest that the proportion of the population that believes in global warming and climate change is declining steadily, and so it may not be long before the politicians lose interest and switch their attention to some other issue, but more on that later.

Reflecting on what has been going on in the world during the past half century; apart from wars and conflicts of one kind or another, there are the obvious trends of technological development and population growth that have led to greater wealth for some, but probably a decline in living standards for those unfortunate enough to have been born in the wrong place and lack the education or political influence to protect their interests. Human greed has always been around, and technology development and the need for raw materials to support the technology have allowed a relatively few individuals and countries, to exploit the situation and become very rich, while the majority have been left behind. Maybe, it is a sense of guilt that has prompted a few rich individuals and countries to offer financial assistance to poor countries in the name of 'development'. However, this financial assistance usually has strings attached and is often used to gain political or commercial influence or personal aggrandisement.

This is the macro-environment within which those foresters who choose to work internationally must operate. Because forests have been steadily reduced and have been relegated to the remoter corners of most countries, a forester's work usually requires visiting these remote areas and meeting many of those unfortunate enough to have been bi-passed by growth and development. Later, we will look at some of the driving forces behind this continued destruction of forests. Despite the fact that the World Forestry Conference in Indonesia in 1978 had the theme of *Forests for the People*, very little was actually done in practice to look at the impact of forestry practices on incomes of rural people and poverty until quite recently. Although the poorest rural dwellers are generally in the areas where there are still forests, the forests are not the cause of the poverty, but the poor communities have rarely been the beneficiaries of the exploitation of the forest resources, more often bearing the cost of the environmental degradation, just as the majority of the population in most oil rich countries rarely benefit very much from the wealth that lies beneath the ground. In some countries, notably Scandinavia, forestry has brought jobs and wealth to those living in the rural areas, as a friend and I experienced during practical work in Sweden. Only in a few countries have forests really benefited the people living in rural areas by providing well paid employment for some and additional income for those fortunate enough to own some forest. In these countries, real poverty in rural areas has been more or less eliminated, though there is still relative poverty due to lack of public services.

Because 'international development' is politically driven, it tends to reflect political cycles in the wealthy western countries that provide most of the funds. These cycles are relatively short-term, since the average life of governments is only four or five years, and if one looks back over the past fifty years it is possible to see how certain topics have dominated development for a few years and then interest has declined as new brooms have come into power with new political priorities. In my view, the media are also, at least partly, responsible, since it has a short attention span and there is a tendency for a topic such as the environment to be flogged to death for a while, and then dropped. Changes in the overall global economy have also had an influence as recession or inflation in the economic cycles have coloured society's priorities

The concept of sustainability has been applied to forest management since long before the politicians found it and introduced it into the politically correct lexicon. However, politicians and businessmen make it almost impossible to practice sustainable forest management, because they want to maximise revenues and profits now, and are not prepared to invest in sustainability. There are trade-offs between what resources we use today and what will be left for the future. If humans were really committed to sustainability, those extracting and selling fossil fuels would be required to re-invest part of the proceeds in an alternative energy source that will produce in the future the same amount of energy that they have extracted.

This would result in a gradual switch from fossil fuels to more sustainable renewable fuels over time and would prolong the life of the remaining fossil fuel resources for future generations. It would increase the cost of the energy now but would promote much more careful and efficient use. However, we have become so used to a high-energy lifestyle that people seem unwilling to accept higher energy costs and politicians are too timid to promote such a concept, so that it is probably no longer possible to implement and enforce such ideas.

Forests are usually referred to as a renewable resource, but this requires that they are protected from encroachment and illegal logging and are either allowed to regenerate naturally or some interventions are undertaken to promote regeneration. This costs money, since some additional expenditure is required to protect and tend the forest, which reduces the net income in the short-term while no one can guarantee that the income stream will continue into the future. The so-called 'invisible hand' of the market does not work for forests, especially for the genetically rich and diverse tropical forests. In fact many economists recommend that forests should be 'liquidated' as quickly as possible in order to realise the capital value and invest it in more profitable ventures. This argument, however treats forest in the same way as mineral resources, and takes no account of the ecological, environmental or social impacts of the loss of forest. By the time that environmental problems such as landslides and flooding, the extinction of a species due to habitat loss, or frequent droughts due to loss of tree cover, have become serious, it is almost certainly too late or too costly to restore the forest. It is very difficult to know for sure how much of the cost of such disasters can really be attributed to the loss of forest and so no one has ever really attempted to measure the economic impact, though nowadays attempts are being made to assess the value of those forests that remain in terms of these broad economic benefits, and to try to get beneficiaries to pay. Thus there are now a few examples of conservationists raising funds to pay for protecting areas of forests to safeguard habitats and of users of water for hydro-electricity generation and irrigation, paying to manage the watershed to minimise the risks of erosion and flash floods.

In Europe, most forests had been destroyed by the end of the nineteenth century, since when, replanting has resulted in forest areas increasing again, but these planted forests tend to be mainly stocked with just one or two tree species and so have less species diversity than the natural forests that were destroyed. Think how many animals and birds that once roamed the forests of Europe are now extinct? However, the temperate forests that covered Europe and North America have much less species diversity than the tropical forests, and most of the tree species have been retained in small woodlands and hedgerows so that given time and careful management areas that are planted or protected may come to resemble natural forests again, though many now have exotic tree species imported from other parts of the world but do not have all the associated fauna and flora.

In the tropics, where species diversity is so much higher, man-made forests can never replace the natural forest with their multitude of habitats for thousands of species. The "invisible hand" may result in more planted forests as wood becomes more scarce and therefore more pricey, but it won't bring back species that no longer have a suitable habitat in which to live, because the plants and animals don't pay rent for the use of the forest as their home and mankind is not prepared to allow them to live rent free. Often the only benefit that the communities living in or near forests in tropical countries derive from their forests are what tends to be called 'non timber forest products', which may include food, medicines and materials for home and household use. As forests are depleted, first by logging and then encroachment and clearance for agriculture, the supply of these commodities, essential to the local communities diminishes and poverty is increased.

Most humans have long since, severed their connections with the forest and much of folklore suggests that forests are dark and evil places. In the rich world, the popular 'forests' for modern man's recreation have generally been tamed, with paths, interpretation centres, signs, bridges, toilets, car parks and other facilities which turn the forests into 'parks' and even in the tropics the current trend for promoting 'ecotourism' is often aimed at better-off people who need certain standards of comfort and safety. In most tropical countries, people living near forests, who practice agriculture, as opposed to hunting and gathering, are frightened of the forest because it harbours wild animals that can often emerge and damage or consume their crops, and may even kill someone occasionally. Shifting cultivation has the advantage of taming the 'jungle' and converting it to a smaller and more manageable secondary forest, with much less troublesome wildlife.

Without air travel, working as an international forester would not have been possible. The availability of satellite observations has enabled forest resources to be measured and monitored more accurately and frequently, which should have resulted in better management, but foresters are not the only people with access to the data. Greedy businessmen and corrupt government officials wanting to make a quick buck from cutting trees and selling the logs can also see where there is still good forest to be cut. Improvements in sea-borne freight have enabled cheap manufactured goods to be shipped around the world, and this also includes logs and wood products, which can be cut in one country, processed in a second and sold in several others. Many of the wood products available in shops in the UK are made in China or Vietnam from wood imported from Brazil and West Africa and frequently illegally imported from the neighbouring Laos and Cambodia as we will see later. People in the wealthy developed countries, and the growing middle classes in many developing countries want more and more material goods, including many things made from wood, most of which can be supplied from tropical timbers. This certainly provides some employment in the tropical developing countries, but generally low paid, dangerous and with very little possibility for wealth creation. Well-connected businessmen and politicians usually capture the wealth in the forest, as we shall see later in the book.

A study that I was involved in in 1991 suggested that tropical countries that allowed log exports, captured less than ten per cent of the total value of the products eventually sold, and even those countries that required all processing to be done in their country, only allowing export of final products, just captured about thirty five per cent. This comes about in part, because taxes, such as import duty and VAT levied by the importing country are based on the high value of the final product. A seventeen per cent VAT on a product worth around US$2,000 per cubic metreof wood can be more than was paid for the original log.It is also because the efficiency of the processing in most developing countries is generally very poor and wasteful compared with most developed countries with the result that almost twice as much wood may be required to produce a particular product.

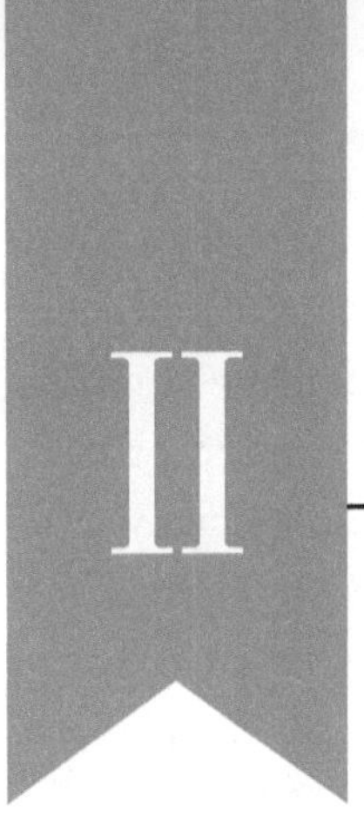

UK Forestry Research (1960 - 1971)

In late 1959, I was fortunate enough to be offered a job by the Forestry Commission, and received a letter telling me to report to the research station at Alice Holt near Farnham in Surrey. On arrival I found that I was the first 'new boy' for some time, and all the researchers were engaged in lines of research that they had been pursuing for some years. In the meantime many new problems had come to light and needed investigation, so that I was landed with a large port-folio of topics to study. One of these was to find out how to reduce the amount of damage caused by high winds that blew trees over before they had reached the optimum size for harvesting. The phenomenon is referred to as 'Windblow'.

Wind Tunnels and Swing-Wings

By the late 1950s some of the plantations established by the Forestry Commission during the 1930s were reaching a size where they were becoming susceptible to being blown over by windstorms. In 1953, an especially severe storm that caused widespread damage around the shores of the North Sea had also flattened large areas of forests throughout the north east of Scotland. The Forestry Commission urgently needed some guidance on the scale of the problem and measures to minimise damage in the future. The potential financial losses if plantations are blown down and have to be harvested prematurely can be very great, especially if the quantity of timber blown down exceeds the normal annual harvest and results in prices being depressed. The challenge was, how best to investigate the problem to get answers quickly. The traditional approach to forest research at that time was to establish trial plots to compare treatments using statistically sound designs, so that valid results could be obtained. The problem was that such trials generally took many years to produce results because the response to the treatments often requires many years growth before significant differences can be measured.Other treatments produce a short-term response, but the differences fade over time. In the case of experiments to investigate possible techniques to reduce the risk of trees being blown over, there is the added complication of not knowing when and where the next big storm would occur and in what direction the wind will come from.

It therefore seemed that the best way to study how trees and forests responded to wind loads of varying strengths was to measure wind loads directly and also measure the forces required to uproot trees under different conditions. The Forestry Commission research station was located near Farnham in Surrey.The Royal Aircraft Establishment at Farnborough was a near neighbour, and it so happened that the RAE had a very large twenty-four-foot (eight metre) diameter wind tunnel. The scientists at RAE were intrigued when they received a request to test trees in the wind-tunnel, and so they agreed to allow the use of the tunnel for two weeks. The National Physical Laboratory at Teddington, in the grounds of which host the tree from which the apple fell on Sir Isaac Newton, was at that time one of the leading institutions researching aerodynamics, and they also assigned a scientist to help with the experiments.

About thirty trees, each about thirty feet (ten metres) tall of five different species were selected and felled in nearby forests and transported to RAE. One by one they were mounted in a tube on an instrumented base and subjected to a range of wind speeds so that the drag on the tree at each wind speed could be measured. The results were astonishing. Man-made structures such as - towers, buildings, bridges and aircraft - experience drag increasing with the square of the velocity; that is double the wind speed gives four times the drag and treble the velocity gives nine times the drag. But trees were different; the drag only increased in direct proportion to the wind speed, so double the wind speed only doubled the drag. Nature was clearly ahead of mankind when it came to aerodynamics. When this conclusion was challenged by the famous physicist Howard Penman (who developed Penman's Formula for calculating evapo-transpiration in 1948) at a Meteorological Society meeting later I had to point out that trees had been experiencing wind loads far longer than technological man and so had evolved to deal with them. The results were real enough, and were a result of the tree branches and the needles or leaves aligning themselves with the wind, more or less parallel with the airflow as the wind increased. As a result the frontal area presented to the wind was reduced. The more the wind blew the more streamlined the trees became. At about that time aircraft designers who were thinking along the same lines came up with the idea of the 'swing-wing' which was intended to have a similar effect for aircraft, and allow them to fly faster.

One outcome from the interest of aircraft engineers in swing-wing trees was that I was exposed to work being done to revive interest and investment in airships. The availability of new fabrics and helium gas meant that airships could be much safer and a number of prototype airships were being tested. It was possible to envisage a number of potential uses for airships in forestry, and we will return to some of them later.

Figure 1. A tree in the 24 ft. diameter wind tunnel at RAE Farnborough for measurements of drag forces

Using the results of the wind-tunnel tests on actual trees, we were able to construct model trees. These had wire mesh crowns and stems made from brass rods that had similar aerodynamic properties to real trees at the sort of wind speeds that did serious damage in forests, around fifty knots. We then assembled forests with these models that replicated the appearance of many forest plantations in exposed sites, with shorter trees on the exposed margins, giving a slight streamlined effect. The model forest could also be arranged to have gaps inside, to simulate roads and small clearings. Sample trees throughout the model forest had strain gauges attached to them so that we could measure the drag on trees at different positions in the forest. This gave us a very good insight as to how wind forces were distributed within forests.

Later we were able to verify the results from the model forests with measurements in a real forest in Northumberland, using anemometers and other instruments attached at different heights to thirty metre tall ex-army signal masts that stuck out above a ten metre tall forest plantation. We also developed a technique for pulling trees over using a horizontal force applied with a winch that was instrumented to give us the strength of the force being applied[1]. Several hundred trees of many species in many locations, grown under a wide range of conditions, (such as on ploughed or drained ground), and at various spacings were pulled over with this technique, and this, combined with the aerodynamic studies, enabled us to work out where, when and under what conditions parts of forests would be at risk of being blown over.

Another finding that came later, related to the swaying of trees. Many people held the view that the wind eddies were synchronised with the natural swaying of the trees in a similar way to the infamous Tacoma Bridge in the Washington State in the United States of America, so that the tree roots were gradually loosened. We then carried out tests on the swaying of trees, by pulling them a certain distance and then releasing them, and found that the natural sway period for the trees was quite long, but the number of oscillations until the tree came to rest was small. The period of wind eddies that contained most of the energy was much, much shorter than that of the trees, so that the effect was the exact opposite - the

---

[1] For anyone interested in more information on this work the Forestry Commission Bulletin No. 40 by Fraser and Gardiner 1967, gives full details.

wind eddies actually helping to reduce the swaying of the trees. Again it seemed that evolution had enabled trees to adapt to the environment to minimise the risk of being damaged by windstorms.

High Explosions in Canada

A year or two after the wind tunnel experiments, the Royal Aircraft Establishment at Farnborough, where they had been conducted, decided to have an open day, and the RAE staff thought it would be interesting and photogenic to have some trees in the large wind tunnel, and so arrangements were put in place to have some trees delivered and the opportunity was taken to do some supplementary studies. As anticipated, the trees caused a lot of interest. Several pictures and articles appeared in the national press and *TheNew Scientist.* Some time later, I received a phone call from an army major who asked if he could come to see me to talk about the wind-tunnel studies. He was from the Royal Engineers, with whom I had served during national service and so the discussion eventually turned to the possibility that I might take part in a secret experiment to be conducted in Canada later that year. He said he would check my security clearance, and if everything was in order I would be invited to a meeting in London in the near future. This was certainly not conventional forest research, but as it was for national defence I was allowed to take part.

At the meeting in London, I was introduced to a couple of scientists from the Royal Ordinance Establishment who explained the reason for requesting my involvement. As part of NATO defence plans for Europe, it was assumed at that time, that should the Soviet Union decide to invade Europe, it would launch a ground offensive with tanks, through Poland and East Germany. Much of the land through which they would have to pass was forested, and the military planners wished to know whether tactical nuclear weapons fired with short-range missiles, could create enough of an obstacle to slow their progress by blowing up areas of forests. Operation Woodpecker was planned to take place in a forest near Hinton in Alberta, Canada where the Abitibi Pulp and Paper Company had logging operations in forest of similar size to that found in eastern Europe. The idea was to explode fifty tonnes of TNT to simulate a small nuclear bomb in an area of forest that had been carefully measured and instrumented and filled with dummies of various sorts. Tanks would traverse the forest before and after the explosion to determine how much they might be delayed.

My tasks were twofold; first to help the scientists develop and test a mathematical model of a tree that could predict how it would respond to varying degrees of blast pressure, and second, to make an inventory of the forest to asses the biomass present to be used to estimate how the forest as a whole absorbed the blast. This preparatory work prior to the blast took about two months. Staying in Hinton with a local family and travelling daily to the test site about twenty kilometres inside the forest became a routine. Sometimes I was given a US Military six wheel truck to drive, and on other days I used a Mini car that a friend living in Edmonton had rashly lent to me. Near the test site was a camp where we had lunch and refreshments, and a field laboratory where we prepared the instruments for attaching to sample trees. A number of trees at different distances from ground zero were selected and fitted with strain gauges at different heights above the ground to measure the deflection. This involved climbing each of the trees with a 'tree bicycle' and suspending oneself with a rope to have both hands free for the work.

Figure 2. Deflection tests on one of the instrumented trees for calibration

Eventually all preparations were complete, and we were allowed to inspect the huge beehive shaped pile of TNT before everyone retired to a bunker at a safe distance to watch the blast. The countdown was broadcast through a speaker in the bunker and then we saw the blast cloud rising above the forest, followed several seconds later by the noise of the explosion. The cloud rose like a dome at first, and gradually turned into a mushroom shape as the dome rose leaving a column of smoke beneath it. When everything had settled down, the site was declared safe, and we returned to assess the damage. At ground zero there was a hole about ten metres deep and about thirty metres in diameter and no sign of any trees. The trees had also disappeared from a zone about another ten metres wide beyond the hole. Outside that came a zone, where the tree stems were just broken at various heights and the shattered bases, one to about three metres tall stuck out of the ground like porcupine spikes, and the top parts of the trees were lying around in jumbled heaps. Further out the number of broken trees decreased and a majority were blown over, and then at about a hundred metres from ground zero the trees were still standing, and just stripped of some twigs and needles.

Dummy soldiers and stores near the zone where trees were shattered were badly damaged, but those in the outer zones where trees had just been blown over had little damage. Tanks were driven through the forest and could make no headway in the areas where trees were shattered or blown over. In the undisturbed forest they could make progress by carefully weaving about to push over some of the smaller trees and pass between the bigger trees. The trees surprised everyone with how much of the blast they had absorbed, when pressure gauges at different distances from ground zero inside and outside the forest were compared. The tree model proved very useful later for studying the effects of storms on trees.

Modelling Growth or How Much Carbon Dioxide can Trees Turn into Cellulose?

During the 1960s computing power was increasing and electronic instruments for measuring microclimate such as wind speed and direction, temperature, solar radiation and humidity were replacing cumbersome mechanical ones. It became fashionable to develop models of biological production systems and forests in several parts of the world were filled with towers supporting arrays of instruments for measuring everything. All this data had to be digested and organised and a number of the models were developed with the aim of predicting how a particular ecosystem might develop and respond to various interventions. One of the main interests was in primary production, or how efficiently plants used solar energy to convert carbon dioxide and water to carbohydrates.

During this period evidence of rising carbon dioxide concentrations in the atmosphere was appearing, and the initial reaction was to try to assess the impact of this on primary production since it should enable trees and plants to have higher rates of photosynthesis. Research started to look at sunlight, and what happened to it when it reached earth. How much was reflected back into space, how much was used in photosynthesis and how much was converted to heat energy. The results were very interesting and underpin the current efforts to model how global warming will change the climate and the impact on ecosystems and on human society. In forestry we began to look at theoretical levels of production, assuming that the efficiency with which the trees used the energy in sunlight to photosynthesise would determine the maximum growth and hence production, which would be limited by the available water supply, nutrients and the temperature regime. A forest manager could manipulate the first two, but the third was mainly determined by the latitude and altitude at which the trees were to be grown.

Theoretical estimates of the growth rates that could be possible seemed unbelievable at the time compared with those being achieved in practice, but subsequent work on tree breeding and development of clones in various parts of the world have shown that, indeed, very large gains in productivity are possible. Until now only a few commercial companies seeking to produce wood fibre for pulp and paper at the lowest possible cost have used such information. However, as the world population continues to grow alarmingly it will become increasingly important to try to maximise productivity of tree plantations in order to minimise the amount of natural forest that must be cleared for conversion to plantations. Unfortunately, the high initial investment involved deters such investment at the present. Before any confirmation of the theoretical calculations became available however, another link with the aircraft industry developed.

At about the same time there was a debate among aircraft designers as to the most suitable fuel for air transport for the future since concerns were growing about long-term supplies of fossil fuels. It seemed that there were two main schools of thought, one favouring switching to hydrogen and the other sticking to hydrocarbon fuels and seeking alternative sources. The then Director of the Royal Aircraft Establishment was of the latter opinion and was also a very keen practitioner of hydroponics (growing plants in a greenhouse in a nutrient rich solution of water) in his spare time. There was also work going on into the technology to convert cellulose to a hydrocarbon fuel by the process of hydrogenation, similar to the way in which margarine was produced. He had the idea that it might be possible to achieve sufficiently high levels of productivity with tree crops to produce enough fuel for the aircraft industry within a reasonable land area. Theoretical calculations suggested that he was probably right, but the Forestry Commission thought it a crazy and impractical idea, and would not take the matter further. Needless to say within a year or so new oil reserves were found, especially in the North Sea, and the price of oil dropped and the idea was consigned to the scrapbook; for a while anyway.

"Woodman Spare that Tree"

In the early days of the Forestry Commission between World Wars I and II, an exotic species called Japanese larch, was found to grow fast on a range of sites and was extensively planted, especially in Wales. One place where it was used a lot was in the grassy hills of the Welsh coal valleys of Glamorgan, around Merthyr Tydfil. These open hills were evidently a favourite place for miners to run their dogs and shoot rabbits, and the Commission came up against much opposition when they fenced off large areas and planted trees. There were records of regular fires suspected of having been deliberately started, but the trees proved to be remarkably tolerant and eventually survived to make a forest. The larch is a deciduous tree and turns a lovely pinkish golden brown in autumn before the needles fall, and having a relatively light and open crown, species like brambles can survive under the trees, and these make a good harvest in the autumn. Thus, in time the locals got used to the trees and found that the shelter they provided made walking dogs much more enjoyable in the bleak winter days, and so when news got around that some of the trees were to be felled, there was an outcry.

Conducting research into the best way to deal with larch crops when they were ready to be harvested was one of my responsibilities, and we wanted to compare a number of ways in which this might be done. The possibilities ranged from clear cutting the larch and replanting the area with either larch again or some other species, to harvesting various proportions of the trees and leaving different numbers of the best ones to grow really big, while planting a new crop underneath. Because of the light crowns many species were expected to be able to grow and thrive under the shelter of the larch, and the system was closer to the way things would happen in nature. However, for the sake of scientific rigour it was essential to clear fell some plots for comparison, and these had to be large enough to reproduce the conditions that would exist if it were done as a commercial operation. The local community, by now thirty years later was mainly a different generation from the one that had resisted the tree planting in the first place, and they had grown up with the trees, so that removing the trees for them was as big a catastrophe as planting them had been for the previous generation. The day was saved for us when the local schoolmaster came to see us in the forest to complain on behalf of the community, but he was kind enough to listen to our explanation of what we were doing, and why, and eventually he was convinced that we were not clearing the whole hillside, and that in time it was likely to become more attractive to look at and walk in, if there were a bigger variety of species and with a mixture of big tall trees and younger ones underneath. He later brought parties of school children to see what we were doing and was able to introduce them to some basic ecology. As a result of his interest, the experiment went ahead without local opposition and goes to show that consultation with the local communities is an important aspect of good forest management. Harvesting a tree crop has a much bigger impact on the landscape than harvesting a field of wheat or potatoes which people are used to happening each year anyway..

Figure 3. View of Michaelston forest from across the valley, showing the larch forest and the plots created to test the growth of trees planted under different crown densities

Recently, in the year 2011, the government proposed to sell off areas of forests established and managed by the Forestry Commission, and public pressure against such a move has been so great that they have had to drop the idea. For those foresters who faced strong opposition to much of the tree planting from the 1960s to the 1980s it reflects a remarkable change in public attitudes towards forests. This is partly due to a generational change as described above, partly a result of the maturity of the forests due to their age and growth, so that they are no longer 'serried ranks' of dark conifers and partly due to the response of the Forestry Commission who have encouraged public access and encouraged a wide range of recreational activities within the forests. Former colleagues tell me that the revenues from public use of the forests now exceed those from timber sales. What a difference fifty years makes! not long by forestry standards

III The Rise and Fall of Sustainable Management of Tropical Forests (1975 onwards)

The massive inflation in the prices of goods and raw materials that took place in the early 1970s following a war in the middle-east, started people worrying that the world was running out of resources and plenty of books were written and conferences convened. Among other things this sparked an interest in the management of tropical forests, which were still seen as one of the remaining world stocks of timber that could be renewed if managed properly. As a result 'development' attention began to focus on management of forests rather than exploitation and ways in which forests could be used to improve life for the communities that lived in or near forests. The theme of the World Forestry Conference changed from *Forestry for economic development* in 1972 to *Forestry for the people"* in 1978, reflecting the change in attitude and a new flavour.

Riots and Railways in Suriname

This first part of this story is more about Suriname than the sustainable management of tropical forests, but the reason for being in Suriname was to work out how to improve the management of their forests and that will come later. This had followed a stint in Yugoslavia doing the same for the temperate forests there. Sitting on the beach one day in Montenegro, in 1973, when it was still part of Yugoslavia, we had some bananas bought at a local shop. These particular bananas had a sticker on them saying 'Bananas from Suriname'. At the time I had not heard of Suriname and had to look in an atlas to find that it was formerly called Dutch Guiana and is in South America. It was therefore a surprise a few weeks later, when the Food and Agriculture Organisation (a United nations agency) asked me if I could go to Suriname for six months. A few weeks later we arrived in Paramaribo, the capital and checked into a local hotel. After a week or so we found a house to rent, and our land rover arrived by sea safely from the UK. Paramaribo is an interesting town, with most buildings in the original part constructed with wood and the largest wooden cathedral in the Americas.

The population of Suriname is a microcosm of the world, comprised of native Amerindians, so called 'Bush Negros' who are descendants of early African slaves who escaped into the 'bush', before slavery was abolished, Asian Indians and others from south Asia, brought in later by the British as 'indentured labour' after the abolition of slavery to work the sugar plantations, and then especially in Suriname, Javanese brought over by the Dutch when they took over the country from the British. There is also a significant proportion of Creoles, who are of mixed blood, mainly African mixed with European or Asian and a more modest number of other ethnic groups such as Europeans, Japanese, Chinese, and Lebanese. Each of the ethnic groups tends to practice the religion of the country from which their ancestors originated, so that Hindus, Muslims, Christians and Jews and a few Buddhists co-exist alongside the animists and other traditional or tribal religions.

A local club with a swimming pool, became the focus of daily social activities, and there we met a British couple who worked for the United Fruit Company. One day I happened to mention the Suriname bananas in Yugoslavia, and the immediate response was "Oh, I can tell you all about that". It seems that shortly after a ship, with a cargo of Suriname bananas, had left Paramaribo bound for the UK, a strike had been declared in Liverpool docks, where the ship was heading. Since the shelf life of bananas is short and the timing of the harvest is precisely calculated so that the fruit can ripen slowly during the voyage and be perfectly ripe on arrival, it was imperative to find an alternative market for the shipment. The company's luck was in as they found a buyer in Yugoslavia and diverted the ship to Bar in the south near the border with Albania, the nearest port in the Adriatic, in Montenegro. Hence we had Suriname bananas on the beach.

Suriname, in the early years of the twentieth century had profitable gold mines and there still existed a section of a narrow gauge railway running from Paramaribo southwards to a point where it met the Suriname river. Originally the railway had crossed the river and extended to the border with French Guiana to the east. The river had subsequently been dammed for a hydroelectric plant to provide power for a large aluminium smelter, creating a huge lake and cutting off the railway. There were stories of a gold mine manager who lived in a mansion alongside the railway, and had a Rolls Royce motorcar fitted with railway wheels inside its road wheels, so that it could travel on the railway into town, and then drive around town like a normal car. On one train journey we got off near the mine manager's house and found it still more or less intact but abandoned. In front of the house a large area had been largely cleared of trees and there was some evidence that it had once been a golf course. In, what had probably been a garage was the rusting rear end of an old motorcar that may have been part of the old Rolls Royce, but it was too far gone to tell.

On another trip we explored a fort that had been constructed by the Dutch in the nineteenth Century to protect the community from what were referred to as the 'Bush negros'. These were slaves and their families that had escaped from the plantations and fled into the jungle, where enough of them eventually congregated to form their own society based on their African traditions. The story goes that in time their community grew sufficiently large and strong, that they were able to raid the settlements along the coast to take crops and other necessities that they could not produce themselves. As ex slaves they were likely to be shot, if caught and so they tended to raid at night and also stole arms. In response to an increase in such raids the Dutch constructed a barrier through the forest along the southern boundary of the area settled. This took the form of a wooden palisade with watchtowers at intervals where soldiers were stationed. The forest was cleared on the jungle side of the barrier, so that raiders had no cover, and could be shot at from the towers. Little now remains of this fortification, but it is possible to find the sites of some of the towers, and at the one we found there was a huge midden beside it where all the old bottles, china, drinking vessels, clay pipes and other rubbish had evidently been tossed. Among the debris I found an old hand made square gin bottle, which is a treasured souvenir of an interesting excursion.

A train ran along the railway line about twice a week taking supplies to small communities along the track and so it was possible to negotiate for the train to stop anywhere to allow exploration of the forest that existed most of the way along the track. The problem was the return journey, because the timing of the train did not conform to a regular schedule, and so there was a strong likelihood of a long wait beside the track. This was solved by getting a pump trolley hitched to the back of the train on the way out, and leaving it beside the track where we got off the train, giving us the means to return relatively quickly. When it came time to return with four people on the trolley and all pumping we timed ourselves at 40km per hour, substantially faster than the train.

The object of one expedition was to search for an old gold mine that had closed in the 1920s.and had since been overrun by the jungle. After trekking through the jungle for about an hour or so, we began to see evidence of some old industrial activity. First, a few sections of railway track that had evidently been a branch line from the surviving track that we had come up in the train, then some rusty wagons, what appeared to be a workshop of some kind, and finally we came across a large water body with the rusting remains of a bucket dredger on a barge resting forlornly at the water's edge. The buckets were about a metre square and a metre deep and were attached to massive chains that would have pulled them around an inclined track, tilted at about forty five degrees so that they would have dredged earth from under the barge and deposited it into another container behind the barge. Trees were now growing out of soil in some of the buckets and the whole contraption was almost hidden by vines that grew all over it. The mine had apparently been just abandoned fairly rapidly, presumably when the price of gold dropped to a level that made the operation unprofitable sometime in the 1920s and it was clear that left alone the forest would rapidly grow back, since just fifty years after the mine was abandoned the area was again thick forest, with trees big enough to harvest. This was the first of many times when it became clear that harvesting some trees does not necessarily lead to the demise of the forest.

Figure 4: Water stop on the Suriname narrow gauge railway

After exploring the site for a couple of hours, we returned to the railway line and mounted the pump trolley. As we pumped away and sped down the line I was reminded of the film "The Titmarsh Thunderbolt" about an old railway line in the South of England threatened with closure, and the scene when two railwaymen disappeared into the fog on a pump trolley singing the *Eton boating song*.

Figure 5. Remains of old bucket dredger in abandoned gold mine in Suriname

At the time of our stay in Suriname, the country had not yet achieved independence from the Netherlands, though it had a fair degree of autonomy. The local government was composed of a number of political parties, which were formed more or less along ethnic lines. Thus the Creoles dominated one party while Asian Indians and Javanese had separate parties. The Creoles tended to work for the government, while the Indians and smaller ethnic groups were the traders and shopkeepers and the Javanese were predominantly farmers and plantation estate workers. The Creoles were trying to increase their influence in the government and were also pushing for independence from the Netherlands, which was generally opposed or not supported by other parties. To push their agenda the Creole leadership called a strike, which virtually paralysed the government for about six weeks. The forest department office was a large building overlooking the river just north of the town centre, and next to it was the tax office. Getting to the office became tricky as almost every day there were crowds demonstrating around the town centre and Governor's Palace, and it was difficult to know the best route to take to arrive unscathed. One day, having dodged some riots and arriving at the office only a little late, it was surprising to see that the tax office had disappeared. The rioters had burnt it down during the night, and being made of wood, it was soon consumed once the fire caught hold, along with all the tax records. The unrest subsided after that for while and the strike ended, but within five years the country had become independent.

Figure 6. Pump trolley on the narrow gauge railway in Suriname

The forests of Suriname are of considerable interest, because of the range of site conditions where they are found. The coastal strip is comprised of alluvial deposits from the Amazon river, which are carried north and then westwards up the coast by the current. All the rivers have sand bars at the mouth and make an 'S' bend as they enter the sea, because the current and the silt deposits push the river flow westwards. The forests on these alluvial deposits are classed as Swamp forests and are more or less permanently waterlogged. The main species found in the forest is called *Virola* and it is ideal for plywood. Until about 1970, most of the exploitation was in the swamp forest and the company Bruynzeel operated a large plywood mill that exported most of its production to Europe. Because the ground was waterlogged, logs had to be floated out of the forest and in some areas, the concession holders were using dynamite to create canals that could be used to float logs to the nearest river. No one investigated the ecological consequences of this practice. From the river landing logs were loaded onto barges for a two to three day journey to the mill. By 1973 the supplies of *Virola* in Suriname were running out, and the company was importing logs from Brazil, to keep its factories running- so much for 'sustainable management'.

Inland from the swamp forests are the marsh forests, which are seasonally waterlogged, and contain different species from those in the swamp. During the dry season the ground could be traversed with tractors, and many farmers earned some supplementary income by dragging trees out for concessionaires. Needless to say they had little regard for the forest and did enormous damage to the remaining trees and left the ground all churned up.

Most of the population is concentrated in a belt across the country from the borders with French Guiana in the east and Guyana in the west on land that was formerly marsh forest. Further inland from the marshy areas the land rises and becomes very sandy; because it is formed by old raised beaches that are found all round the Caribbean sea. The natural forest is an open savannah type with grasses and shrubs under an open tree canopy. The native Amerindians were mainly settled in these areas and they burnt the grass regularly to improve the grazing for wild animals that they hunted. The Forest Service was not happy to see such scrubby forests and had invested large sums of money to establish plantations of a tropical pine that was native in Honduras. The plantations were moderately successful, but there were not enough of them to supply a modern industry, and so there were questions as to what should be done with them.

Further inland again, beyond the savannah, was what the Forest Service called 'dryland forest'. This was a tropical moist forest that contained a number of deciduous species, because the annual rainfall is seasonal and there is a significant dry season. The government had defined a 'Productive Forest Belt' running east-west across the country covering the forest that could be readily accessed from the navigable sections of the main rivers. The Governor had the powers to allocate concessions within this forest, and the Forest Service had only an advisory capacity with no powers over the size of the concessions or the way in which they were managed, despite the fact that a five year development plan in 1967 had committed to bringing out legislation to empower the Forest Service. Most of the concessions had been allocated to foreign companies who were primarily interested in short-term profit and so exploited the forest in a fairly haphazard manner cutting just accessible trees of a limited number of species. One characteristic of this dryland forest, which also applies to much of the Amazon forest, is that a high proportion of the species have very dense and hard wood so that they do not float, and are therefore more costly to transport and process. However, the wood of several of the species including the well-known *Greenheart* is very durable, and they therefore are of special interest for uses such as jetties and railway sleepers, which are exposed to the weather. As long ago as 1973 the People's Republic of China, as it was referred to in those days, had established trade with Guyana and Suriname to purchase mixed hardwoods, and they were much less choosy on species than the European market.

From around 1970 until about 1980, by which time Suriname had become independent, Dutch scientists were conducting a wide range of research into the management of the dryland forest to ensure their sustainability, but political instability during the 1980s put a stop to much of the work, and to attempts by the Forest Service to improve forest management. For a short time during the energy crisis of the late 1970s, charcoal became big business and large areas of forest were cleared to make charcoal. The method used was very primitive earth kilns, which involved digging trenches about ten metres long, two metres wide and two metres deep and then filling them with wood. The wood was then covered over with the soil removed from the trench except at a few places where access was left to allow fires to be lit. Once the fires were well established they were also covered over and the wood was then allowed to cook slowly for a week or more. Once the process was considered to be completed the earth was tamped down to seal it completely and everything was allowed to cool for a few days. Once cool enough, the pit was opened up and the charcoal removed. Much of the labour was imported from Haiti, as there were not enough Surinamers to do the work. After the price of crude oil slumped in the mid1980s interest in charcoal around the world declined and the operations in Suriname ground to a halt.

There was also a Dutch company working in Suriname to establish oil palm plantations, and they were having difficulties in disposing of all the debris from the trees in the forest that they were clearing. All the commercially interesting species had already been removed and so they were left with large quantities of unsalable logs and branches. Since the company was a general trading company they were also interested in the possibility of making charcoal from the residues but they did not want to use the very environmentally damaging and inefficient earth kilns being used by the government. At that time I was also involved in a company in Scotland making charcoal, using small steel kilns that could be moved from site to site, and these had been made to our design by a blacksmith in Port Seton. He was delighted when I came to him with an order for fifty kilns for export to Suriname, which he duly fulfilled. Later on a visit to Suriname I was interested to see these kilns in operation and the company was very happy that it could turn what had been a costly nuisance into a good revenue stream.

Emergency in Thailand and First Attempt at Sustainable Management of Asia's Forests

In early 1975, work in the FAO Regional office for the Asia-Pacific in Bangkok was aiming to improve the management of the tropical forests in the region. Much of the region's forests were being rapidly opened up to provide timber for the growing national economies, especially Japan and Korea, which were both developing rapidly at that time. After initial visits to all the countries in the region and discussions with the forestry departments and other experts working in the field, it was decided that the best way forward was to try to support forestry departments in the region to establish some pilot forest management units, intended to demonstrate at a commercial scale, how to manage the natural forests in the region on a sustainable basis. The Royal Forest Department in Thailand had agreed to establish one of these pilot units in the southern part of the country where there were still, at that time extensive areas of pristine forest, and had proposed two possible sites. It was agreed that a couple of senior Thai foresters and I should make a reconnaissance visit to the proposed sites, in order to get some basic information to determine which site was the more suitable, and to start the planning process. Accordingly the three of us boarded a train one evening in Bangkok and travelled south all night, arriving in the provincial city of Chumphon the following morning. We were picked up by a local departmental vehicle and driven to the airport, where a small plane was waiting for us. The intention was to fly west for about a hundred and fifty kilometres and over fly the first of the two sites, and then turn south and cross the mountains to Hat Yai where we would overnight before flying over the second site the following day.

Everything went according to plan at first, and we had a good view of the first site, which was beautiful virgin forest covering a broad basin and rising up the slopes of the surrounding mountain range. It appeared to have high biodiversity value, as well as a range of site conditions, including flat land, wetlands, and land on a range of slopes, that would be good for demonstrating how to practice multiple-use and sustainable forest management. We then turned south heading for Hat Yai and began to encounter heavy cloud. The pilot called Hat Yai on the radio, which advised flying further west to skirt round the mountains and keep in clearer weather. After a while it became clear that the clouds were thickening and not dispersing as we had been advised, and so the pilot decided to turn around and fly east and hope to reach the sea and fly down the coast clear of the mountains.

However, the weather was clearly worsening and the wind was increasing, and so he then decided it was better to land at Surat Thani where there was a small airport, but he couldn't establish radio contact with it. As we approached the reason became clear as the airport was under reconstruction, and there were bulldozers and trucks all over the runway. However, by this time fuel was getting low and the imminent storm made landing imperative, so the pilot flew low over the runway to try to indicate to the workers on the ground that we needed to land. They all looked at the plane, some waving, but clearly did not get the message, so the pilot made a further circuit and then lined up for his approach. Only when we were almost on the ground did some workers realise our intention and a couple of drivers rushed to their trucks and just managed to move them enough in time so that we could avoid hitting them.

As we taxied to a parking spot near the terminal, the heavens opened and we were obliged to remain inside the plane, except one of the Thai foresters gallantly volunteered to make a dash for the terminal and try to find a vehicle to take us somewhere. After an hour or so he returned to say that he had found a vehicle to take us to Nakhon Si Thammarat, about a hundred and fifty kilometres to the south, and half way to Hat Yai, from where we should be able to catch a train for the rest of the journey. The rain eased a little, and the vehicle arrived, and we set off. What no one had mentioned was that the road to Nakhon Si Thammarat was not paved, and in places was almost impassable because of the heavy rain. To add to the difficulties, an election was in progress in Thailand at the time following the riots of 1973, and several parties had election convoys travelling in one direction or the other along the same road. It was therefore quite late at night, around ten pm before we finally arrived at Nakhon Si Thammarat. We called on the home of the local forest officer, to be told that there were riots in the town; crowds were at that moment trying to burn down the governor's house and we should get out of town as soon as possible. The reason turned out to be that supplies, that had been sent down from Bangkok for people who had suffered from recent flooding, had somehow disappeared and the intended beneficiaries, not having received anything, were rioting to try to get hold of the supplies intended for them.

The railway station was fifteen or so kilometres away to the west of the town, and so we made our way there. We were told that a train was due in an hour or so, and so we had time for some refreshments. At around three am the train eventually arrived, having been delayed by floods following the storm that had caused our troubles. We boarded it, and managed to find some space to stretch out a little, as all the sleeping berths were occupied, and it was not long before I must have fallen asleep, because some time later I was awakened by noises like gunfire, to find that the train had stopped. It turned out that the train had some wagons attached to it also carrying supplies intended for flood victims. The news had evidently got through to local people, who had managed to stop the train to get their hands on the supplies. Although at the time, such acts were not seen in a religious context, the people in the area are predominantly Muslim, and evidently felt that they were being treated unfairly. Since that time the conflicts between the majority in the region, who practice Islam, and the government have increased and become a serious armed conflict.

Eventually the train was allowed to depart, when the contents of the wagons had been unloaded, and we crept into Hat Yai just as dawn was breaking. The remainder of the trip was relatively uneventful, but later that same year when in Hat Yai again another similar incident occurred. I had gone to a bank to change some money, and had completed the paperwork and was waiting for the Thai Baht that I needed, when I became aware of a commotion on my right. Looking round I found myself standing next to a man pointing a gun at the cashiers. In those days the bank tills were quite open and the man was reaching over the counter and scooping up bundles of notes and stuffing them into a bag. It was a few moments before it really dawned on me or the other bank staff what was really going on, but I soon became aware that one of the clerks sitting at a table behind the till had produced a gun from under his table and pointed it at the thief just to my right. He saw it at the same moment as I did, and fortunately for me turned and fled before the clerk could fire. The clerk followed him out of the door, but he had leapt onto a motorbike waiting outside with an accomplice ready for a rapid getaway, and they disappeared in seconds. My first thought was whether the bank would complete my transaction, and also I was worried that I might become embroiled as a witness, so I asked the cashier if they had enough cash left to pay me, and was mighty relieved when the answer was "yes". I managed to get my cash and leave before the police appeared. Such incidents subsequently became commonplace in that part of the country.

In late 1976, this work ground to a shuddering halt as the UN found that it was bankrupt following the dramatic period of inflation resulting from the 'oil-shock' of 1973, which had, among other things led to contributions to the development fund from rich countries drying up, and so all projects that had not yet been fully approved were stopped. The project that I had been working on had been endorsed by eight governments in the region, but Malaysia, where one of the pilot management units was to be established, was still having discussions between the Federal and State governments over details of the project, and had not yet signed by the time of the deadline; so the project was cancelled. Over fifteen years later the circle was completed and an opportunity arose to take up the work again in Indonesia, as we shall see in a later chapter.

IV Forest Products (1973-85)

s the world recovered from the austerity of the second world war, such matters as poverty, environment and biodiversity were very low on the political agenda, and foresters who tried to promote sustainable management or conservation were more or less ignored. World demand for timber and paper was growing rapidly and production came first. Globally, the focus in forestry was on providing industrial raw material and the industries to convert logs to products. The rise of the Japanese economy and the Vietnam war, stimulated strong demand in the far east, while the recovering European economies did likewise, mainly in West Africa. Tropical developing countries with forest resources were keen to exploit them. Much of the World Bank's and the UN Food and Agriculture Organisation's efforts were directed at feasibility studies for pulpmills and sawmills and developing countries were encouraged to establish industrial plantations. This was partly driven by the rising prices of most commodities due to inflation, which for a while made such investment look commercially attractive.

Forest Management in Greece

In the early 1970s, Greece was run by a military dictatorship and was still sufficiently poor to qualify for technical assistance from FAO. Although the holiday maker to Greece, who visits mainly beaches and the classical architectural sites such as Delphi and Olympia, might think that Greece has few forests, and what they have is mainly the Aleppo pine and scattered cypress trees found in the hills around Athens, this is a false impression. Travel a few kilometres up into the mountains and you will come across some very fine forests of Black or Corsican pine with Spruce, Fir and Beech in the central and southern regions and Oak and other hardwoods in the north along the border with Bulgaria.

Management of these forests was introduced by the Germans before the first world war and followed a sound but very conservative principle. Each forest unit had a series of permanent sample plots scattered through it that were re-measured every ten years. Comparing two successive measurements of the timber volume in all the plots gave the total growth that has taken place during the ten year period since the previous measurement, and this was used to determine the Allowable Annual Harvest (AAH), which is set at about seventy five per cent of the growth, to allow for losses due to disease, fire and other natural causes.

This process should ensure that the total growing stock in the forest remains more or less constant in the long term, and hence the production is sustainable. However, in Greece, and also in Yugoslavia and Turkey, where the same principles were applied, it was not working in practice. The successive measurements of the sample plots were showing that the total growing stock was declining and so the managers reduced the Allowable Annual Harvest to less than seventy five per cent, because they thought that they had been overcutting. Uncontrolled grazing, mainly by goats, which outnumbered humans several-fold was a great concern to the foresters who were reluctant to fell older trees for fear that young regeneration would be eaten. It turned out on close investigation, that the field managers had been correctly measuring the growth and setting the AAH at seventy five per cent, and then they were passing the information to their District chiefs for approval, who, it turned out, were reducing the AAH still further to be 'on the safe side' and then passing their plans on to the Regional chiefs for approval. The Regional chiefs did the same and then passed it on to headquarters in Athens for approval, which also did the same. As a result the AAH that was eventually approved by headquarters was less than fifty per cent of the growth.

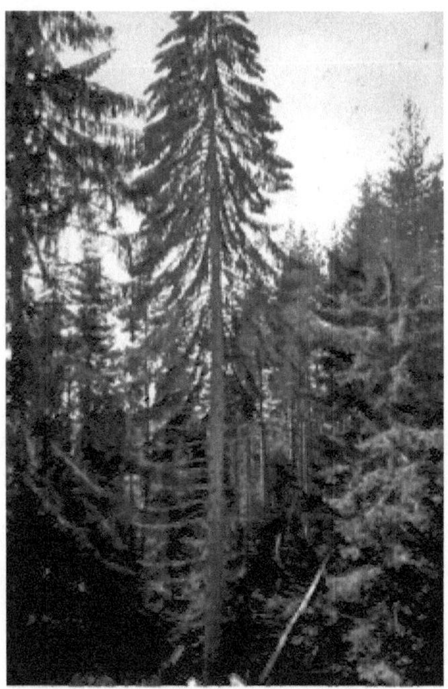

Figure 7. Fine, spruce-pine fir forest in central Greece

This should have resulted in the growing stock increasing over time, (like leaving capital and most of the interest in the bank), but because the remaining trees were ageing their growth was past its peak and was slowing down, with the result that the growing stock was actually declining. By the early 1970s the system had been in place for about sixty years and the forests had almost stagnated, as it was mainly composed of very old trees that were dying rather than growing. It took some effort to persuade the Greek foresters of the reality of the situation and to agree to a temporary increase of the AAH to about a hundred and thirty per cent of the growth, so that the old mature trees would be gradually replaced with young trees that would have much higher growth. This did involve keeping goats out of the forest, to give the young trees that would rapidly fill gaps where old trees had been felled, time to grow enough to survive being nibbled.

In the early 1970s the Greek economy was beginning to grow and imports of wood and wood products were rising fast and accounted for about three quarters of total consumption. Apart from one small pulpmill and a small factory making chipboard, the wood processing industry was comprised of small and generally old-fashioned sawmills serving local markets. Several were powered by steam engines using all the residues as fuel. The government was therefore keen to have a large-scale modern industry, to increase domestic supplies of timber, save foreign exchange and create more employment in rural areas where farming is very marginal and does not provide adequate income. The Greeks delight in the story of the farmer, who went to work in his field one morning, and before starting work, put his jacket down and went to a nearby stream for a drink. On returning from the stream, he could not find his field and spent the rest of the morning searching for it. At lunchtime he decided to go home and on picking up his jacket found that his field had been underneath it all the time.

The government had therefore asked FAO to supervise an inventory of the forests in the mountains around Karpenision in order to determine if there was enough raw material for a new integrated wood industry, primarily to produce pulp and paper and possibly particleboard. The inventory certainly showed that there was plenty of timber, but because the trees were mostly quite old and therefore large, it did not make economic sense to use them for making pulp since they were far more valuable for sawing to make a range of useful products. The idea for a pulpmill was therefore dropped and in due course a feasibility study was conducted that confirmed the gains to be had from using the timber for joinery products and furniture.

Timber Trend Studies

In order to attract investment into forest industries and plantations and to assist national forest policy decision makers, FAO began researching trends in the supply and demand for forest products in the recent past, and projecting these into the future in order to show how demand was likely to increase for different product categories. The first European Timber Trend Study (ETTS I) was published in 1953, and it was revised and updated more or less every decade after that so that the fifth in the series published in 1996 was looking at trends for the twenty first century. But it wasn't only FAO that was looking at future supplies of timber globally as there was a real concern that the world was running out of resources, and in both the USA and in the UK several similar studies were published by commercial forecasters and government agencies. Needless to say they all came up with similar findings, because they were all using the same basic data.

By the mid 1990s modelling of national and global timber supply and demand had become a major field of study and more than a dozen different models were in use around the world, with such esoteric names as HOPSY, HUGIN, FASOM, TAMM and FOLPI. The huge increase in computing power had enabled more and more sophisticated modelling to be undertaken. While this was good for the modellers who developed scenarios for the developed countries, where forest management is generally sound and there is reliable data, nothing was done in developing countries, where such forecasts of what is likely to happen in the future could be really useful in anticipating the likely outcome of continuing current practices. Three main reasons for this lack of effort in developing countries were: lack of experienced and skilled modellers in most developing countries; a lack of will on the part of those responsible for forests who failed to see, or intentionally overlooked the need for such information, since making money from cutting down the forests was more important; and the lack of, or unreliability, of the data.

The main source of global information on forestry and stocks of timber is held by FAO, but the quality is very variable, since they depend heavily on governments both to supply the data and also to approve what they publish. In many countries the forestry departments have been reporting false information, either because there is widespread illegal logging and they don't have a true picture, or because, especially in autocratic countries, they have been reporting old figures for so long that they can no longer admit to the real situation for fear of reprisals. Thus many countries, especially tropical developing countries have been blissfully ignoring the real state of affairs. One of the ruses used to avoid reality is to reclassify what is forest from time to time, so that figures for inventories at different times cannot be compared and the situation looks better than it really is.

As described later in Chapter IX illegal logging is widespread in many tropical developing countries, sometimes supplying up to half the domestic requirement for wood, so that real consumption may be as much as double that estimated by using the official figures on production. In the Philippines and Thailand, more or less all the forests were logged both with official sanction and illegally in the three decades from the early 1960s to the late 1980s so that by the end of the period both countries became net importers of wood and wood products. The total volume of logs that was reported as having been harvested and exported or processed into different products over the thirty year period was less than half the volume that was lost from the forest over the period according to the official inventory figures based on changes in forest area and average growing stock. Presumably some of the difference was due to unreported use for various purposes, and some was logged illegally and so not reported, and some must have gone up in smoke or just rotted away following clearance and conversion to agriculture. Together, the unaccounted loss of timber plus all the branches and non commercial species would have contributed huge amounts of carbon dioxide to the atmosphere. At the time this aspect of forest clearance was not something that people thought or cared about very much, but it has been and still is going on throughout most of the tropical forest belt and has now become the flavour of the month as people wake up to the fact that maybe as much as twenty per cent of all carbon dioxide emissions has been coming from the destruction of forests. Thankfully in other parts of the world, especially Europe, forest areas are expanding and the trees are growing and re-cycling some of the carbon dioxide.

In the early 1990s FAO provided technical support to Indonesia with World Bank funding, to carry out a national forest inventory. The results were published in 1996 in a summary national volume and with separate volumes with more detail for each of the major islands. The Ministry of Forestry had previously published official statistics on forest cover classified according to the function, that is, according to whether it could be used for production, or needed to be kept for protection of the environment or for wildlife conservation and national parks, and it was a surprise to find that the areas given in the new national Forest Inventory for the different forest categories were almost identical with those in the official statistics produced about ten years earlier. At the time when the inventory was published I was actingas Coordinator for a large forest management programme funded by the British government through the Department for International Development, and we were providing policy advice to the Ministry of Forestry on ways to improve forest management. Reliable information on the condition of the forests was essential in order to provide sound advice. We therefore examined the results of the inventory for the individual islands in detail and aggregated them to compare with the national results. To our surprise the result showed a very different picture from that in the national report. The aggregate of the areas in the island reports was only about seventy per cent of those in the national report. According to one of the international team that had been advising on the inventory the Forest Department had pressurised FAO to make the national results 'consistent' with their official statistics and hoped that no one would compare them with the aggregate island data.

The British Government had provided support to Indonesia during the first half of the 1980s to carry out a major national land-use planning study to assist the Department for Transmigration to choose areas outside Java, suitable for establishing new settlements for transmigrants. The study used early LANDSAT images and mapped the land use across all the so-called 'outer islands', that is all the islands other than Java.. Although the Forest Department had been given a copy of the final report, because it was intended for transmigration it had just been 'filed' and was never used for forest planning purposes. However, it provided a very valuable base-line for comparing with the national Forest Inventory to study changes in land cover during decade from mid 1980s to mid 1990s. We come back to this later when examining illegal logging in Indonesia. (see chapter IX)

Feasibility Studies

During the three years from 1976-79, feasibility studies for new wood processing capacity were the order of the day, taking us to Indonesia, Greece and back to Suriname. A commercial company funded the first and development funds channelled through banks funded the other two. All resulted from previous work in the countries. In Indonesia, a British company had acquired a stake in a concession on an island to the west of Sumatra, through the takeover of another company during the great boom in mergers and acquisitions of the mid 1970s. Since the company had acquired the concession as part of a package, they had little expertise in tropical forestry or forest management and had hired a lawyer, an accountant and an estate agent in Singapore to run the operations. Little wonder that it was having problems! The London headquarters could not understand why the company was not making a profit, and they were under increasing pressure from the Indonesian government to submit plans for their investment in wood processing, which was one of the conditions for the concession.

The first task was to establish the quantity and quality of the timber resources available within the concession. The crew in Singapore had hired an American 'logging expert' from the Philippines to do an inventory of the forest and to design and layout a road system. He had negotiated a contract whereby he was paid according to the volume measured and the length of road surveyed and constructed, which gave him a big incentive to find as big a volume as possible of wood and build as many roads as possible. Needless to say, his data showed that the volumes were very large and that the further you went into the island, hence more roads needed, the bigger the volumes became. He had just walked off with around US\$ 6 million in fees, which explained why the company was not making a profit. He escaped to New Zealand, and although the company considered suing him, I believe they never did.

When we checked the inventory, we found that he had used tables intended for calculating the volume of felled logs to calculate the volumes of standing trees. The log volume tables take the mid diameter of the log and assume it is more or less a cylinder to get the volume, and builds in an allowance for the bark. Volume tables for standing trees cannot use the mid diameter, unless someone climbs the tree, and so normal practice is to measure the diameter near the ground, and the tables make adjustments for the taper, so the tree is treated as a very tall truncated cone. The log tables for a mid diameter of say about fifty centimetres will give a volume about twenty five per cent larger than a standing tree volume table for tree with the same diameter at the base. So our American friend's estimates of the volumes were about twenty five per cent higher than actual for which he was handsomely paid. One of the dangers of not using a professional forester for what appears to be straightforward work, and this is just one of many examples of companies investing heavily on the basis of little or no or misleading information. This includes Banks who are particularly prone to sanction loans to forestry operations without consulting with impartial professionals, which may be why so many have gone bust in 2008/9. Its one form of toxic asset! Many British banks and investment funds invested heavily in the Indonesian pulp and paper companies that went bust in 1997/8 following the Asian financial crisis- will they never learn?

By comparison the other studies in Greece and Suriname were fairly straightforward, and the former resulted in the design for a processing plant that was eventually built and did quite well, while the other could have done the same had it not been for a *coup d' état* and a long period of political strife in the country

Consequences of Inflation: Shipping Beech Wood from Scotland to Hong Kong

At about the same time as the Indonesian feasibility study, my consultancy company decided to take a stand at a trade exhibition in Kuala Lumpur to promote itself in the region. We had a small stand that attracted an encouraging amount of interest and plenty of visitors. One of the visitors was a Chinese gentleman from Hong Kong, who among other interests represented a British company called Rabone Chesterman that produced a range of measuring instruments, including wooden rulers. At the time these rulers were made from wood from the Box tree, which was hard, close grained, stable and could be worked to a fine finish, ideal for rulers. The supplies of wood came from China, and we had an interesting discussion about supplies and likely future price trends. About two weeks after returning home to Scotland, we received a telex from our new Chinese friend in Hong Kong to say that he had a client looking for Beech wood from Europe, which they wanted to try as a substitute for Ramin from Indonesia and Malaysia, which was becoming more scarce, difficult to procure and very expensive.

We examined the possibility of procuring Beech from Europe, especially France, Germany and Rumania, which is known for the pinkish colour of the wood. All the sources we tried were not interested in a small order, and with inflation still rampant they quoted high prices, and so we approached a small saw-miller in Scotland. He was happy to oblige and offered a container load of sawn Beech at a very competitive price. In those days a Scottish shipping company, the Ben Line had regular container sailings to Hong Kong, and so we were able to organise the shipment directly to Hong Kong within two weeks, and a deal was done.

Figure 8. Loading a container with Scottish Beech for shipment to Hong Kong at the former Hardengreen sawmill

Not long after the shipment arrived in Hong Kong, I was in the region and arranged to stopover in Hong Kong for a couple of days and meet the buyer to see what the wood was being used for and how suitable it was. I met with our Chinese friend and he drove me to the factory that had bought the wood, somewhere in the middle of the New Territories. The factory was a big shed, maybe about one hundred and fifty metres long and twenty metres wide, and inside it were rows of machines for sawing, cross cutting, planing, drilling, sanding, polishing and so on, and the whole place was filled with a fine dust from sanding machines. Each machine had two or three girls operating it, and in between the machines were still more girls, stacking, moving and sorting the pieces coming out of the various machines.

The products emerging from this production line were tennis bat handles, mug stands, toggles for duffle coats and other clothes, clothes pegs, toothpicks and a whole range of similar small products. When I worked out the number of finished items that were obtained from our container load of wood and the price that the factory was selling them at, it appeared that the small factory in Hong Kong was increasing the value of the wood about twenty times, despite the fact that the wood had come from half way round the world. The importance of adding value post harvest, and who gains the benefit from adding that value is a thought that has remained with me ever since that visit, and has surfaced time and time again over the years. It also shows that one doesn't need to have timber raw material available down the road, as enterprise and imagination can offset the cost of bringing wood half way round the world. Mind you, part of the equation would be the very low wages and lack of attention to health and safety that minimised the costs of production. Had the Hong Kong factory paid wages comparable to those in Europe and invested in goggles, earmuffs and dust extraction equipment it would probably have been a somewhat different story

Wood Processing in Japan

Japan is a major importer of wood from all around the Pacific basin. Some countries that have forest resources, such as Malaysia and Indonesia have managed to build up processing industries, and export finished products, that have a higher value than logs, but the smaller countries such as Papua New Guinea and Solomon Islands have neither the financial or human resources to do likewise. The Solomon Islands were worried that they were losing out and they wanted to study options for increasing their revenue from log exports. Most of the logs exported were going to Japan, and so as part of the study it was necessary to find out what happens to the logs once they reach Japan. With few trained staff, the Solomon Islands were unable to carry out detailed inspection of shiploads of logs, to determine the sizes and species of the logs, so that most were exported as mixed species and so called 'super smalls' which were logs smaller than the normal minimum diameter of fifty centimetres.

A visit was arranged to the harbour in Osaka, where a boatload of logs was being unloaded. This required a ride in the famous 'Bullet train' from Tokyo, and a first view of the densely populated and built up coastal belt between the two cities as we sped along. At the harbour, the logs were being lifted out of the ship's hold and placed into piles on the dockside according to species and size. A small but important proportion of the logs were very valuable hardwoods such as *Rosewood* and *Vitex*, a very durable hardwood not unlike Teak, and I learnt that these were sold at a high price for making veneer, which was later sold on to furniture factories. The species were grouped by colour, strength, hardness and durability and each group was then sold to factories specialising in products requiring particular types of wood. The middleman who had imported the logs had almost tripled their value in this way - a good business for him, but a big loss for the Solomon Islanders.

Figure 9. Sorting cargoes of imported mixed logs in Osaka harbour according to species and quality to maximise utilisation

Later, a visit was made to one of the furniture factories that had bought some of the logs to see what they were used for. The one visited was medium sized and made a range of furniture, including beds, chairs, tables, wardrobes and sideboards. The wood was first sawn into boards and stacked for kiln drying. The kilns used waste wood and sawdust as fuel. Once dried, it was taken to a large workshop equipped with a number of saws and planers. As it was sawn into standard sizes, any defects, such as knots or splits were marked, and the pieces so identified were sent to a further saw where the defects were cut out with the minimum waste of wood. The resulting short lengths were then sent to a further line where they were finger jointed to give longer lengths of defect free wood. After planing and sanding the wood was cut, drilled and fashioned according to templates ready for being assembled into furniture. All the waste, consisting of the defects that had been cut out, shavings, sawdust and drilling waste went to the kiln, so that a hundred per cent of the wood was utilised. The factory was perfectly clean. This compared with a factory in the Solomon Islands, and others seen in many developing countries later, where whole planks of wood with small defects were discarded and off-cuts, sawdust and other waste lay in great heaps all over the place, so that only about fifty per cent of the wood actually finished up in products. No wonder the Solomon Islands could not compete!

Figure 10: Typical sawmill in developing country with waste everywhere

Different Perspectives of the Pulp and Paper Industry

Although chronologically the following stories about the pulp and paper industry occurred over a wide timespan and do not fit exactly into the period covered by this chapter, they provide some insight into what the sector looked like in the 1970s and 80s since when the sector has grown steadily and expanded throughout the world over the past three decades dominated by a relatively small number of very large companies that can take advantage of massive economies of scale. Thus the modern industry as described in the next Chapter and again in Chapter IX is quite different from those covered in the following stories

Pulping in Sweden

As a student of forestry it is essential to get some practical experience during the university vacations. In 1955 an opportunity came up to work with the major Swedish forest products company called Svenska Cellulosa, in their forests around Sundsvall. Getting there on limited funds meant hitch hiking from Scotland for as much of the journey as possible. Another friend was also offered a job, and so the two of us set off to get to Harwich for a ferry to Hook of Holland. From there we crossed the Netherlands and northern Germany to catch the train from Lübeck that ran north to cross the straits by ferry to Laaland Island in Denmark and on to Copenhagen. Wearing a kilt proved particularly valuable for getting lifts, as whether out of curiosity or out of love of the Scots, drivers seemed very inclined to stop and offer us a lift. Many even went out of their way to help us on our journey. A day or so to see Copenhagen and the famed Tivoli Gardens was a must before the crossing to Malmö in Sweden. From there we managed to get to Stockholm, but had to resort to train for the journey further north to the small town of Ånge where the branch line to Sundsvall leaves the main line to the north and the SCA had the office for the forest operations where we were to be employed.

Working in the forest was hard going, mainly felling trees, de-branching and de-barking and crosscutting into two and a half metre lengths and then stacking the logs in triangles so that the gaps between the logs aided drying. Home was a small log cabin, shared with about six other workers, including a couple of Finns and a local lady came in morning and evening to cook meals. The mosquitoes were really troublesome, and the only way to keep them at bay was to smoke a pipe to create the biggest possible cloud of smoke. It seemed to be more effective than repellent. Every two weeks we were paid, after which the Finns would disappear for a few days. We learnt that they went to Sundsvall and drank all their earnings, only returning to work when their heads had cleared. Towards the end of our time, we asked if we could work in the pulpmill for a week or so to experience the process. We got a message that our request had been granted, and so we boarded a train for Sundsvall.

Arriving at the mill, we were shown to a hostel where we could stay and were told to report for work as soon as possible. It turned out that we had been assigned to a special trouble-shooting team that had to deal with any breakdowns. The plant had been built in 1902 and so at the age of more than fifty years was getting rather decrepit, but it was evidently cheaper to keep it going than to close it and build a new one. Our first job was to clear a huge hopper full of wood chips, situated above one of the large digester, which took the form of tall cylinders with an hatch at the top which was opened for filling with chips and then closed and sealed during the cooking process. The digester needed repair, and the chips had to be moved to an adjacent one so that it could start immediately and not wait for its hopper to fill. We worked on a floor built just below the tops of the digesters so that it was very hot, and there was a long row of digester tops sticking up about a metre through it, and in the ceiling above each digester was the opening from a hopper that was full of wood chips concealed by the ceiling. Our work involved opening the hatch from the hopper above the digester that was to be prepared and allowing a few tonnes of chips to spill out on the floor, and then barrowing it to the other digester that needed filling. Each digester holds about fifty tonnes of chips, so it took almost a whole day to transfer the full load, especially as we had to keep going up to the top of the hopper and climbing in to move more chips down to the opening. Because we kept stopping the flow to avoid covering the whole of the floor at the level of the digester openings, the chips lost momentum and refused to come out when we opened the hatch again.

The next job was almost as strange. After digestion, the pulp is drained from the digester into a washing room, with pipes running from floor to ceiling that had nozzles on them. As the pulp enters, the pipes rotate and the nozzles spray water to wash it. During the night, one of the pipes had broken, fallen over and been buried by the mass of pulp, and we had to recover it so that it could be repaired. This involved one of us being lowered through a hatch in the top of the tank and filling large buckets with soggy pulp, which the other lifted out of the tank and transferred to a neighbouring tank. Having removed enough pulp to expose the top of the broken pipe, we started to dig down to find the broken section. After digging for a few hours and having a hole about two metres deep, we asked the foreman how much further to go, to which he replied "about another six metres". It took us the rest of the day and half the night to reach the broken pipe, whereupon the engineers came to fix it and we were allowed to go back to the hostel for a rest.

Several more similar jobs kept us fully occupied for the rest of our time at the mill, including clearing logjams at the chippers and moving chips from conveyors that needed repair. When the time came to go home, we went to collect our pay, and the manager handed us our pay packet with a smile, and thanked us for our efforts. We could hardly believe our eyes when we opened the pay packet outside, as it contained more money than we could have earned in almost a year working in a forest in Scotland, so we really enjoyed the trip home by train, which we could easily afford.

Paper Mills in Greece

Many years later, a pulpmill figured again in a curious story. Greece has forest areas that are used for timber production, mainly in the higher parts of the Pindus Mountains, and along the northern border with Bulgaria. As mentioned earlier these forests have been managed very conservatively for many years and the possibility of an increased harvest of logs from the forest meant more raw material could be available for processing, and so it was necessary to inspect the existing facilities before recommending the most appropriate developments. One of the factories visited was a pulp and paper mill at Patras. This mill was making Kraft paper, but had difficulty in obtaining enough raw material to operate at its capacity. The management had heard that a boat full of bagged cement had run aground in the Adriatic, and they purchased the cargo to recycle the old bags. The ship was eventually re-floated and towed to the harbour, and during our visit, truckloads of bags of cement were being brought into the factory yard. There, a group of men were opening the bags, emptying the contents in a big pile and throwing the empty bags onto a conveyor where they would be taken to a shredder for feeding into the pulping process. One of the products coming off the production line was cement bags, and these were being used to re-bag the cement coming out of the recycled bags. I never did work out the logic of all this recycling, but I assume that the cement, once in new bags, could be sold for a higher price than it could if left in the old bags, but whether the cement was any good and the extra income covered the cost of re-pulping the bags remains a mystery.

Paper Mills in India.

Paper mills come in all shapes and sizes and some are running out of raw material, while others never had enough. With four tonnes of wood needed to produce a tonne of pulp for papermaking, raw material supplies are a perennial problem. During the late 1970s and early1980s some plantations of *Eucalyptus grandis* were established in the southern Indian State of Kerala. As these plantations approached maturity a feasibility study for a newsprint mill was undertaken, which relied on the plantation manager's records for information on raw material supply, and a decision was made to go ahead and invest. However, there were two small problems. First, some of the plantations were suffering from diseases, especially a bark canker called 'pink disease', (see next chapter) and second a substantial proportion of the supposed plantations did not exist. The reason the latter situation had arisen was because regular fires burnt parts of the plantations, and when the areas were replanted they were treated as new areas, so that double counting was taking place and the actual area was only about two thirds of the reported area. Raw material supply problems were not confined to the Kerala mill and the pulp and paper industry in India had been lobbying the government for some time to allow them access to more land for tree planting, with very little response. The Ministry of Industry therefore requested an inspection of all the major pulp and paper companies to assess their overall raw material supply situation

One of the paper mills to be visited was in Calcutta, and it was arranged with the company that they would pick me up at the airport. A smart white Ambassador car, basically a 1950s British Morris Oxford, with white linen seat covers, curtains over the windows and a smartly dressed chauffeur was waiting at the airport on my arrival, and I was whisked off into the busy streets of the city as dusk fell. After driving for about half an hour, the car turned off a wide main road down a very narrow lane. To my astonishment, the narrowness of the lane was caused by shanties on either side, made from plastic sheeting up against walls about three metres high, each filled with vertical tiers of beds, so that there seemed to be three or four layers of people sleeping or resting and others milling around in front. After a few hundred metres, we came to a very large and solid looking gate in an archway, about three metres high, with double doors about the same overall width. The chauffeur gave a loud hoot on the horn and the gates opened just long enough to let us shoot inside, and they slammed shut immediately behind us. We passed into another world.

Inside the compound, in front of the gate was a large building, and the car turned right, passing between two buildings and emerged into a wide open grassy space. It was difficult to see in the dark, and after a short distance we pulled up in front of a large house. Three or four broad steps led up to a wide veranda with large pillars supporting the roof, where two grandly uniformed staff were standing, who rushed to collect baggage and open doors. Inside was a grand hall with shiny wooden floors and several doors leading off. I was shown through one of these into a bedroom with a large four-poster bed draped with mosquito netting. This was the company guesthouse. I was told that a meal would be served shortly in the dining room across the hall. After a pleasant meal it was time to turn in for the night.

The following morning was bright and sunny and a walk before breakfast to explore the surroundings seemed in order. The grassy space I had seen through the gloom the night before, turned out to be a golf course sloping gently down to the banks of the Hugli river, one of the branches of the Ganges in its delta. There were a few scattered trees, and the whole place had a well-kept air about it. A hundred metres or so across the grass were the buildings I had seen the previous night, which were evidently the factory. After breakfast, a couple of junior managers from the company appeared to show me round the factory. Inside the noise was pretty deafening and the machinery was clearly quite old, as there was steam leaking out of several pipes, and small jets of pulp squirted out in places like snow making little cone shaped heaps on the floor. At the end of the line, rolls of paper were being formed like huge toilet rolls, and as I passed by the paper making machine, I suddenly saw a plaque on the side with the name Bertram Sciennes Ltd. Edinburgh 1867. I used to pass the factory frequently when at home, but had no idea that I would see one of their machines over a hundred years old and still going strong. It seemed that the factory had problems getting fresh raw material and was largely recycling a mixture of papers, including some imported from various countries in the region.

 Plantations as a Solution to Tropical Deforestation(1980-90)

With the demand for paper and other forest products increasing rapidly many companies wanted to build new pulpmills in tropical countries to be as near as possible to the main expanding markets. Some companies built pulpmills to utilise all the small trees that were left in the natural forests after logging, but the big variation in the properties of the woods, some light, some dark, some hard and some soft, made it difficult and expensive to control the pulping process in an optimal way, and this reduced profitability. It therefore became obvious that natural forests were not the most suitable source of raw material and that it would be essential to have plantations of fast-growing species to provide raw material with consistent properties in the future. The only problem is that it costs a lot of money to plant and tend a hectare of trees, and even the fastest growing tree species in the tropics will need six or seven years before they can be harvested. That kind of investment is not very attractive to bankers who like to make a 'fast buck', nor is it something that many small farmers can take on as they do not have enough time to plant and tend the trees as well as grow the food crops they need to survive, and they cannot go without some income until the trees are ready to harvest.

A modern chemical pulpmill requires about four tonnes of wood fibre to produce a tonne of pulp, and the high cost of cleaning the water and other effluents to avoid pollution must be spread over a large volume of pulp to minimise the unit costs. This means that modern chemical pulpmills tend to be very large and require both a huge investment of around $1 billion as well as enormous areas of plantations to keep them supplied. Some pulpmills established in the 1990s in the far-east produce about one million tonnes of pulp annually and need a plantation area of more than 250,000 ha to keep them supplied continuously and this means an investment of at least a further $200 million in the plantations, which must precede the investment in the pulpmill for the trees to be ready for harvest when the mill is ready.

Although the total wood costs are only around thirty per cent of the total manufacturing costs, because of the large volumes of wood required, the unit volume cost is critical since a one dollar increase in the price of a cubic metre of log becomes about four million dollars increase in total manufacturing costs. As a result the price paid for pulp logs has to be low, and this makes the investment in the plantations appear to have a very low return on the capital invested, compared with the investment in the pulpmill. As we will see in the second story, pulp companies are very reluctant to have large areas of plantations on their books and would prefer it if someone else invested in the plantations. Some companies got round the problem by resorting to tree breeding and the Aracruz company in Brazil in particular achieved remarkable results, eventually increasing productivity more than threefold.

The main method for increasing productivity is to first undertake a selection process to identify individual trees of the species to be used that have outstanding growth, quality and form characteristics. These may be from a single species or sometimes a hybrid between two species. These special individuals are then propagated vegetatively via rooted cuttings, so that each of the progeny is genetically identical and has the same growth, quality and form characteristics as the parent tree, referred to as clones. If you walk through most tree plantations you will observe that there is enormous variation between the biggest and the smallest trees. In a plantation established with clones all the trees are almost identical.

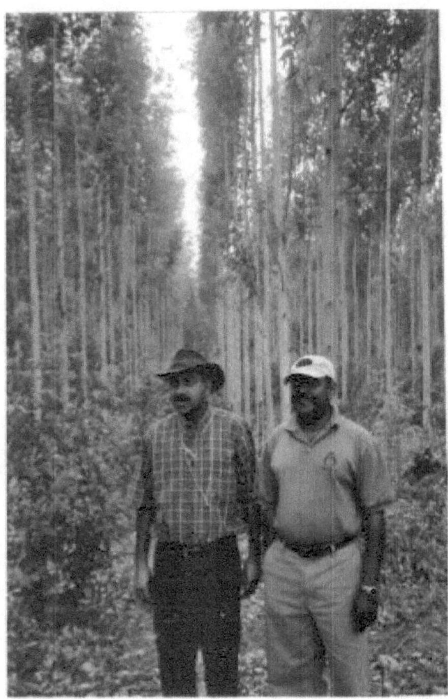

Figure 11: A plantation of clonal *Eucalyptus* in Lao PDR showing how the trees are almost identical

Floating Pulp-Mills and Plantations in Brazil

In Belem near the mouth of the Amazon, a small office issued passes for selected guests to visit the Jari river concession owned by a Mr Ludwig. He is an American millionaire who had made his fortune from the shipping company Universal Containers. He had reportedly purchased a concession of around two million acres (eight thousand square kilometres) and proposed to plant the fast growing species *Gmelina arborea* to provide raw material for a pulp mill. Many stories had emerged about wholesale destruction of forest and plantation failures, and so the opportunity to visit and see what was going on was welcomed. Having been issued with the pass, I was informed that a flight would depart the following morning at 0900 hours. The plane was a Norman-Britten Islander, a small eight-seater twin engined plane, and there were five passengers, including myself, plus the pilot. I felt privileged to have the right hand front seat beside the pilot, as it gave me the best view. We took off to the north and flew out over one of the many channels of the Amazon in its delta. The flight took about an hour and a half, flying over islands clothed in swamp forest dispersed among the various meandering river channels. The airport at the Jari concession was located on a low plateau and as we approached we could see some of the plantations.

A short drive from the airport was the base camp, with houses for permanent staff, a guest house for visitors a club with swimming pool offices and workshops. Having checked in, the first stop was some of the plantations. In some areas the growth of the *Gmelina* was really impressive, but in other areas, growth and form were disappointing. The staff had learnt the hard way that *Gmelina* is sensitive to soil conditions, and where the soils were sandy and the forest had been cleared and burnt, it did not grow well. On heavier soils and where the vegetation had been pushed into heaps and left to rot, the growth was good. On the sandy soils they had found that pines were a better bet than the *Gmelina.* So the stories about plantation failures were partly correct, but the problems had been identified and put right. These problems could probably have been avoided had foresters with experience in the tropics been used at the beginning, rather than using north American temperate foresters with no relevant experience, and then having to bring in specialists later.

The next visit was to the site of the pulp mill. The mill had been built in Japan on a huge barge, and towed across the Indian and Atlantic Oceans, up one of the mouths of the Amazon and then up the Jari river. A huge lagoon had been constructed and large logs driven vertically into the ground as piles, and then cross cut at a standard height. Once ready, the lagoon was opened to the river and flooded so that the barge could be floated in and positioned over the piles. Once the barge was in place, the lagoon was closed off again and the water pumped out so that the barge descended to rest on the piles. A second smaller barge was also brought in, in the same way equipped with all the service facilities such as the power plant. This novel way of constructing and delivering a pulpmill was cheaper and safer than bringing in all the materials and constructing the mill *in situ.*

After the mill was a visit to see the railway track under construction. With the plantations spread out over a huge area, the cost of hauling the wood to the pulp mill by truck is high, because maintaining roads, especially during the rainy season is very difficult and expensive. By using a railway, the distance from plantations to the rail track could be limited, and the railway could operate during all seasons. The whole operation was very impressive, and the idea of a floating pulpmill was to return later, but the lesson was clear; research the factors such as species, soils, and climate before investing large amounts of money. Some years later, Mr Ludwig could no longer sustain the investment and sold out to a consortium of Brazilian banks.

Sometime after returning to Scotland, there was an article in the *Scotsman* newspaper about a Glasgow firm of consulting engineers who had developed concrete floating structures for use in North Sea oil exploration. This seemed to have some potential for floating pulp-mills. A phone call got a positive response and later a meeting was arranged to discuss the details. Within a month or so a concept design was ready, and a preliminary estimate of the likely cost. About that time something came up in Indonesia, and while there, the opportunity was taken to renew contacts with the north American forest products company, Weyerhaeuser Corporation, which had a large concession in eastern Kalimantan. They were logging natural forest, but were also experimenting with plantations with a view to developing a pulp mill at some future date. The idea of having a floating pulpmill was discussed, since the plantations would be scattered over a substantial area, and it looked as if it might be cheaper to bring the mill to the wood rather than the wood to the mill. Indonesia seemed to be a potential market for floating pulp-mills as it has so many islands and an enormous coastline. Back of the envelope calculations suggested that the idea could have potential cost savings.

However, about that time, the Indonesians were beginning to try to enforce agreements with foreign investors to build processing capacity in the country, rather than exporting logs, which brought little benefit to the country. After lengthy discussions most of the foreign companies, including Weyerhaeuser, decided to pull out. This marked the end of the floating pulp mill idea, as well as the beginning of the take over of Indonesia's forest resources by President Suharto's cronies.

Pension Funds in the USA

During the late 1970s it became fashionable among the major North American pulp and paper companies, to invest in 'timberlands' to secure their raw material supplies. Because of environmental concerns about logging in the mountainous terrain in the Pacific North-west, many companies were moving their operations to the south-eastern states of Georgia, Alabama, Tennessee and South Carolina, where abandoned cotton land was available relatively cheaply, and the native pines grew reasonably well. However, the companies tended to transfer the wood from their timberlands to the factory more or less at cost price, which made the production of pulp and paper look very profitable, but made the forests appear to be giving a very poor return on the capital tied up in them. As a result, during the 1980s many of the companies started trying to dispose of their timberlands in order to improve their balance sheets.

At about the same time, in the UK, pension funds had found that young plantations could be purchased relatively cheaply, for little more than the cost of establishing them, due to special tax arrangements that enabled investors to offset the cost of establishing plantations against tax liabilities from other sources of income. These young plantations were at the stage of peak growth, though not yet ready to harvest, so that their book value was increasing rapidly, and this made them an attractive long-term investment, that helped to sustain the value of their investments and provide income at some future date. Because of the limited supplies of suitable plantations in the UK, some of the more adventurous funds wanted to investigate the prospects for investment in the USA.

The prospect of selling their timberland assets to a pension fund, and concluding a management agreement with the fund to look after the plantations and have first refusal of the timber when harvested appealed greatly to the pulp and paper companies. It also looked good for the pension fund too, because they could invest in an appreciating asset with minimal management costs for the fund and a steady income stream in the future. It looked like a marriage made in heaven. Despite the apparently favourable conditions, the UK pension funds considered the risks too high, and did not go ahead with any significant investments. Within a few years, forest plantations were out of favour with pension funds in the UK and most of their holdings were disposed of.

Plantations in Korea

At the end of the Korean war and for a few years thereafter, the population of the country was largely (about seventy per cent) rurally based and depended on agriculture and forestry for their livelihoods. With no national reserves of fossil fuels, wood was the main source of energy used by rural households for heating and cooking and was especially important during the harsh winters. Most of the land is in private ownership with each family having a small land holding, but as a result of the war most of the land that had been tree covered was bare due to the conflict and the heavy demand for wood. The world bank provided a loan to the government to fund a major tree planting programme, and the population showed the same dedication and commitment to planting trees as they would show later to adopting industry and technology. As a result huge areas of tree plantations were established within a decade, and the momentum was maintained for many more years. However, the country was steadily industrialising and people were migrating to the cities so that by the late 1980s the proportions of urban and rural population had reversed, and there was no longer such a high demand for fuelwood. Nor were there sufficient people to tend and harvest the trees.

Although the central government exercises strong control over the economy, and all planning is done by the national Forestry Administration, all forestry activities on private land, which is more than seventy per cent of the total, are supervised by provincial bureaux. In addition there is a strong network of Forestry Associations down to the village level that provide technical support and guidance to individual farmers. Through this arrangement Korea had succeeded in establishing over three million hectares of forest in a twenty-year period, which is almost fifteen per cent of the national land area, a truly remarkable achievement.

Having succeeded in establishing about three million hectares of plantations in less than three decades since the end of the Korean war, the Forest Administration was at a bit of a loss as to what to do with all these plantations. Most of them were privately owned in very small parcels, so that a whole hillside might be covered with trees but the land owned by hundreds of people. With so many of the former inhabitants, or their children, having migrated to the cities the owners were often difficult to trace and there was no organised way for coordinating the management. The country had not quite reached the point at that time when it made the transition from 'developing' to 'industrialised' status, and so FAO still had a project to advise the government on forest management.

In order to assess the situation it was necessary to tour the whole country and meet with the Provincial authorities to find out about the extent and condition of the plantations under their overall supervision. While this was in general a very interesting opportunity to see much of the country and take in some of the historic sites along the way, it also had a serious drawback that would not have affected an ordinary tourist. The tour was organised so that more or less a whole day was spent in each province starting in Seoul and working south down the western side of the country and then across the southern part and back up the eastern side. The first province was entered very near Seoul, and the schedule consisted of a morning visit to some forest or a nursery, a pause for lunch, an afternoon visit and then a dinner with a number of provincial officials, usually in a town near the border with the next province.

Figure 12: 12 year old plantation of pine in Korea used for the cultivation of Shitaki mushrooms grown on stacks of wood under the tree cover.

On most days the dinner was in a hotel where we spent the night, and the following morning we were picked up by officials from the next

Province and followed a similar schedule with them. The drawback was that each dinner was hosted by the provincial government staff who were happy to have the opportunity of food and drinks paid for by the government, and they took full advantage of the situation to order all the local delicacies and enough booze to float a small battleship. The booze was the biggest problem, because it generally consisted of a rice based spirit that was quite strong, and local custom required that each of the hosts, maybe six or seven men, should link arms with the guest in turn and drink a toast for long-life and happiness. This meant that as a guest one would have to drink about six glasses of spirit for every glass drunk by any of the hosts, though in between toasts the balance was somewhat reversed. Fortunately almost all the officials did not seem to have a head for alcohol, so that despite the imbalance in the amounts drunk, they finished up under the table first. After ten days of this it was a welcome relief to get back to Seoul and have a bit of a rest.

One area of particular note that we visited is north of the major port city of Busan in the south-east of the country near a city called Kyongju where an extensive area of Loess soils had become very severely eroded as a result of deforestation. The soil is very soft and easily weathered and the rushing waters after rain had created many large gullies so that a huge area had been almost devoid of vegetation. Apparently the area is right under the flight path for commercial flights coming into Seoul from Japan and elsewhere and had been observed by the president on one occasion when he was returning home. An instruction was issued that the area "will be rehabilitated". Around 1973 a major effort was launched to restore tree cover over about five thousand hectares and the work was completed in four years. Photographs produced for me during the visit showed the rapid transformation as huge gangs of local people were brought in, first to create terraces on the steep slopes and then to plant trees along each of the terraces. The species chosen for the planting was mainly the so-called 'false acacia' or Black locust (*Robinia pseudoacacia*) and this grew very rapidly so that within a couple of years the whole area was tree covered and the erosion had stopped. Ten years later during our visit the trees were six to ten metres tall and were flowering profusely so that bee keeping and honey production had become an important source of income for some local residents.

Figure 13. Seriously eroded land near Busan, Republic of Korea, most of which was restored by terracing and tree planting.

Young seedlings were springing up all over the place and although the terraces could still be seen they were slowly weathering, but with the tree roots and a deep covering of dead leaves and litter meant the soil was well protected

'Pink' Disease in Kerala, India

As much as thirty years ago, the demand for paper in India was expanding rapidly and the industry was regularly complaining about shortages of raw material for making the pulp, required to make paper. In the southern state of Kerala, the government had initiated a plantation programme in the mid 1970s, using *Eucalyptus grandis* that was intended to produce fibre for a pulp mill. The hilly land to the east of the coastal towns of Cochin and Trivandrum is mostly state forest land and had substantial areas of grassland resulting from clearance of forest in the past, which had failed to regenerate as a result of frequent wildfires. The presence of the grass reduced the suitability of the land for agriculture. Despite these difficult conditions the land was considered to be well suited for the growth of the trees, based upon the performance of some old trials where the trees were impressively large after only about 8 years of growth. About eight thousand hectares of the grassland had been allocated to a Kerala Newsprint Project for planting with *Eucalyptus grandis*, and this was considered to be sufficiently extensive to ensure the continuous supply of about seventy per cent of the raw material needed for the pulp and paper mill that was designed to produce about 100,000 tonnes of newsprint annually. The remaining thirty per cent of the raw material was expected to be from a species of reed, which had longer fibres than the *Eucalyptus*, essential for the newsprint. In 1983 the mill began limited production.

The theory of raw material supply seemed perfect, but as so often happens, reality is very different from theory and there were big problems on the horizon. Not least of these was the fact that the reeds had started a mass flowering, after which they die, and so supplies of the reeds could not be guaranteed, but there were also big problems surrounding the potential supply of the wood fibre. The most obvious problem affecting the trees, which the government had identified and responded to, was the outbreak of a fungal disease on the trees called 'Pink disease' that produced large salmon pink coloured cankers on the stems of the trees and eventually killed them. The government had responded to this by establishing a special 'fungus investigation unit' to find ways of controlling and eliminating the disease. Less obvious was the unnoticed problem due to the fact that the actual area of plantations was far less than the records suggested. This had come about because each year, some of the plantations had been burnt by wildfires, and when they were re-planted the following year they had been recorded as new plantations and added to the total area established, thus double counting. In India, as in many countries there is a regular turnover of staff due to promotions and retirements and so a new forest officer would generally accept the records left by his predecessor. Before the days of satellites and GPS it was very difficult to measure accurately the total area of many smallish blocks of plantations scattered over a wide area and so an incoming officer had no real way of knowing how many hectares he was taking over. Although there were maps of the plantations, these showed the area allocated rather than the area planted and did not show the age of the trees in each block, which could be very mixed if parts had been replanted several times.

Using the available records of the area planted each year and making allowances for the proportion that was actually replanting rather than new planting, based on records of fires it was possible to estimate that the actual area of plantations was only about sixty per cent of the recorded. This, combined with the reduce yields as a result of the disease, would have a big impact on the short-term raw material supply situation for the mill. Recommendations were made for completing the planting of the whole area as a matter of urgency, and improving the standard of records to provide accurate information of the rates of growth and the true stock of wood fibre.

It seems that the mill survived all these problems, as it is still operating, but it reports on its website that it is using an unspecified amount of recycled paper as raw material. This may be good for the environment, and perhaps the silver lining to the problems in the forest is the pressure that it has imposed on the industry to recycle its fibre.

Oil Bearing Trees in Vietnam.

I arrived in Ho Chi Minh City, formerly known as Saigon, in mid afternoon, and after checking into my hotel, decided to take a walk round to see what sort of things were being sold in local markets and shops. As I was walking down a narrow pedestrianized street, I became aware of two women that seemed to be stalking me. After a few minutes, one of them came up to me and tried to attract my attention, but also raised my suspicions, which were shown to be justified, when out of the corner of my eye I saw that the other woman was trying to grab my wallet, which was in the breast pocket of my shirt. I reacted instinctively and raised my hand, which was enough to scare her off, and the two of them vanished into the crowd. Later sitting in the hotel dining room for an evening meal, I was mystified by a constant low rumbling noise. It became clear, when I looked out of the window that it was coming from thousands of motorcycles streaming past on the street outside. Most of them had either two boys or two girls on them, and it was clear that this was the local dating arena.

The following morning, my two colleagues arrived, and a minibus organised by the forestry department picked us up and we were driven south towards the Mekong delta. We crossed the main channel of the

Mekong river by ferry near Vinh Long, then another branch of the river to reach Can Tho. From there we turned north west and travelled up the side of the river to Long Xuyen, the provincial city for An Giang Province where we were to stay for the following week.

The purpose of the visit was to advise the local Vietnamese forestry staff on the establishment and management of plantations of a species called *Melaleuca*. Prior to the war with the Americans which started in 1955 and dragged on until 1975, much of the land in the Mekong delta was covered with natural *Melaleuca* forest, which provided good cover for the Viet Cong (supporters of the communist government in North Vietnam). The Americans sprayed much of the forest with Agent Orange or burnt it with napalm, to kill the trees and make it harder for the Viet Cong to hide and move about without being spotted. By the end of the war, *Melaleuca* forests only covered about 16% of the land area of An Giang province, and a similar proportion of other provinces in the delta.

Melaleuca trees have a number of valuable uses, of which firewood is probably the most important, but it also produces a very good aromatic oil that is widely used in decongestant medicines, and is the basis for a small industry distilling the oil. The trees also make very good poles used locally for scaffolding and construction. As a result of all these uses the remaining forest under real threat and our Vietnamese colleagues said that the area covered by the trees by that time (1991) was reduced to only 4,000 hectares or 1.6 per cent of the provincial land area and the government was therefore trying to reverse the trend by restocking areas. The demand for fuelwood was reckoned to be three or four times the sustainable supply from the *Melaleuca* forest, and much of the shortfall was being met by clearing mangrove forests along the coast, with severe negative environmental consequences

The soils in the Mekong delta are variable, and there are substantial areas that have acid sulphate soils that are not suited to rice growing. Many farmers had unwittingly cleared *Melaleuca* forest to grow rice, only for it to fail because of the soil acidity, and it was these areas that the forest department was hoping to restock with trees. They were collecting seed from existing patches of forest and scattering it on the areas to be restocked. In order to improve the growth of the trees, they were constructing canals around each of the blocks to be rehabilitated and making dykes from the soil removed from the canal. The canals were connected to the river, and had sluice gates so that the water level could be controlled. The dykes prevented flooding during the wet season, and the canals could be kept full of water for much of the dry season to keep the soils moist and reduce the risk of fire. Inspection of some of the areas established during the previous five or six years, showed that the quality of the trees was extremely variable, and that all the trees were from only one of the two species that were natural in the region, and the inferior one at that. We were able to make a number of recommendations on the management of the plantations in the future, especially regarding the sourcing of seed, that should help to improve productivity and survival if implemented.

On the last night of our stay, a party was organised by our hosts, with live music and plenty of eats and drinks. The locals were glad to be able to enjoy themselves as it was paid for by the government's entertainment fund. One of the senior staff revealed that he had been a local Viet Cong commander and had fought against the Americans for many years. He was a very good singer and kept us well entertained. All the Vietnamese staff present performed songs and eventually the inevitable came when I was asked to sing. Singing is not my forte and the only song that came to mind was *Auld Lang Syne,* so I hummed the tune, and to my surprise the band new it well and struck up with a rousing version. I did my bit and everyone joined in and it brought the party to a successful close.

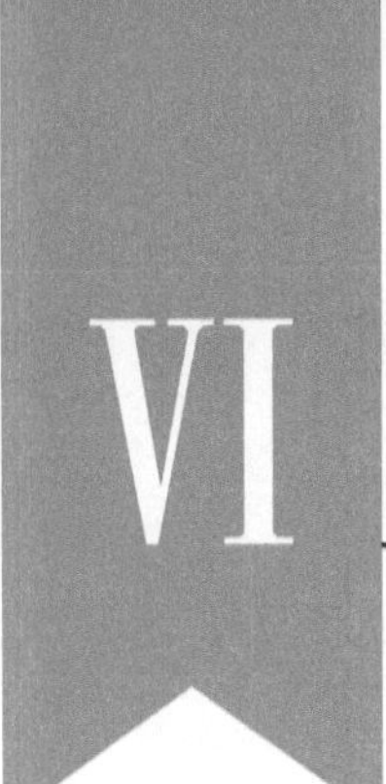

VI. Energy, Agroforestry, Mangroves and Desertification (1976-88)

Charcoal Everywhere and Charcoal-oil Mixtures in Guyana

The period from about 1976 to 1985 was a time of high oil prices and strong interest in renewable energy and alternative sources of energy. Technology developed during the Second World War for gasification of wood and charcoal making was revived and modernised. Wood began to look competitive with oil and coal as a source of energy and the large oil companies investigated the use of charcoal-oil mixtures as a means to utilise a bio-fuel in liquid form that had the advantage of avoiding the need to change the combustion technology while reducing the cost of the fuel. Charcoal was still widely used for steel smelting at that time especially in Brazil and Malaysia, which lacked major coal resources and several African countries wanted to follow suit, including Kenya. Consequently for a few years there was great interest in fuelwood plantations and in charcoal making, but as the price of oil came down, interest declined and eventually ceased more or less completely.

Figure 14. Brick kilns in Minas Gerais, Brazil, used for making charcoal for steel making

In Guyana, in South America, there are huge bauxite mines, which supply a smelter to produce the aluminium. The smelting process requires enormous amounts of electrical energy, which in Guyana was produced by an oil-fired power station. Many countries use hydropower, which though more costly to build does not suffer from fluctuations in the oil price. The industry in Guyana was being crippled by the high oil price and the government requested some assistance from USAID to investigate the possibility of using charcoal oil mixtures. This would of course depend on the local availability of charcoal.

In 1980 the price of oil rose sharply and remained high for the next five years, and this prompted interest in biofuels again. One line of investigation that British Petroleumfollowed was the possibility of making charcoal-oil mixtures. Charcoal is mainly carbon and can be refined into a very fine powder that will mix with oil, so that it functions as a fluid rather than a solid. The energy in a tonne of charcoal is about two thirds of that in a tonne of oil, so that the mixture has slightly less energy than pure oil, but provided the charcoal does not exceed about forty per cent of the total, it can still be used as fuel for oil burning generators, and so avoiding the need to convert to boilers for raising steam for turbines.

With the average price of charcoal at that time in the range of US$ 120-150 per tonne, an oil price of more than US$25-30 per barrel makes charcoal cheaper than oil on an energy equivalent basis. In 1980 the average price of oil, adjusted for inflation to current prices was the equivalent of almost US$100 per barrel, and remained above an actual price of US$30 per barrel until 1986 so that the idea of being able to substitute forty per cent of the oil with much cheaper locally produced charcoal was very attractive. The bauxite mine and smelter were located north of the main rainforests of Guyana in a belt of drier more savannah like forest, where the trees were relatively small and in many ways ideal for making charcoal.

The big problem with industrial-scale charcoal is ensuring the regeneration and sustainability of the forest resource. In the early days of the industrial revolution, charcoal was used widely for iron smelting and this led to the wholesale destruction of huge areas of forest, because charcoal makers were happy to cut down the trees, and did so at a rate far faster than the forest could regenerate, but were not willing to manage the forest to ensure sustainability. Thus in Guyana there was a real danger that promoting charcoal production would result in the rapid disappearance of the forest, despite the fact that good management of the operations could have ensured sustainability.

The area around the bauxite mines and the smelter was sandy, having formerly been beaches during the era when sea levels were higher, and the forest was rather short in stature and not very dense. The species however, had quite dense wood and would have been very suitable for charcoal, but there were strong reasons to doubt the local capacity to manage a large-scale charcoal making operation in an environmentally sound and sustainable manner. There would have been enough wood, provided the most efficient charcoal making technology was applied, but that required substantial investment and training of many people, which seemed to be unrealistic. As a result, the ball was thrown into the long grass by recommending more detailed studies and an environmental assessment, which didn't materialise straight away and was eventually rendered unattractive by the fall in oil prices. This was a case of something being technically and economically feasible, but undesirable from an environmental point of view. Fortunately the price of oil gradually declined and the price of aluminium began to rise, and the operators lost interest in the charcoal idea; but maybe it will come back again with new oil price highs.

The oil price boom in the early 1980s that had prompted interest in charcoal in Guyana had also stimulated the interest In Suriname discussed earlier where staff of the Forest Department saw potential for making charcoal on a large scale for export to Europe and China, which had established arrangements for purchasing logs from the country. However, they were using wood from dense high forest and by using the rather primitive technique described earlier, their activities were both inefficient, unsustainable and environmentally destructive, though better than just burning all the debris from forest clearance.

Figure 15. Large-scale charcoal making in Suriname using earth kilns

In the northern arid provinces of Nigeria, wood is very scarce, but charcoal is the preferred fuel for cooking. However, as supplies of wood dwindle the price of charcoal is rising beyond the reach of many of the poorest people. One solution would be to plant trees in these northern states to provide raw material for the future, but because of the lack of rainfall, the productivity of the trees would be very low, and huge areas would be needed to meet the demand for charcoal. An alternative that was being considered by the forest authorities was to grow trees in the wetter southern province, where yields would be much higher, and transport the charcoal to the north. Charcoal is a bulky commodity, and although railway was one means for transporting the charcoal, the net flow of cargo was from south to north, the same direction as the charcoal had to go, so that there was no chance of getting costs reduced by providing return loads in empty wagons.

Under these circumstances, airships seemed to be a real alternative, because the bulkiness of the load is not a factor. A railway wagon with a capacity of twenty tonnes of normal cargo could only carry about six or seven tonnes of charcoal, and this contributed to the high cost of road or rail transport. However, an airship with a lifting capacity of twenty tonnes, can lift twenty tonnes of charcoal, even though the volume may be about three times that of conventional cargos, since the airship is physically much larger and the cargo is suspended underneath it. The economics looked very interesting, but unfortunately, Nigeria began the civil war around that time and no one would consider a risky investment, but airships have survived, and will re-appear later.

In Scotland, the demand for charcoal, mainly for barbecues was growing rapidly, resulting in rapid growth in imports. Together with some friends we established a small company to produce charcoal locally from sawmill residues and branchwood and other residues from tree felling. It was the time when Dutch Elm disease was rampaging across the country and there were large numbers of dead elms that needed to be disposed of. Turning them into charcoal seemed to make better sense than just burning them. Because the wood to be converted to charcoal was scattered in small quantities all over southeast Scotland, it was decided to use mobile steel kilns. These were both efficient and could be easily moved from place to place on a specially designed trailer.

Figure 16. Charcoal making with portable steel kilns from dead elm in Scotland

For several years the company succeeded in producing a steady flow of good quality charcoal, but as the pound sterling appreciated in value, imports became more competitive. To make matters worse, the biggest retailers, the supermarkets and DIY stores, would leave their ordering until the last minute, just before Easter, and then demand hundreds of tonnes to be delivered to their stores all over the country, and this proved an insuperable problem for a small company. Smaller retailers across Scotland became regular customers and even sold small quantities during the winter, but this segment of the market was too small to make the operations commercially viable. It seemed that in order to grow the company a very large investment would be needed to make the transition from an artisan to an industrial operation. In addition to the large amount of capital required investigation showed that it would not be possible to obtain sufficient quantities of wood for a sustainable operation.

Fuelwood

The primary source of energy for people living in rural areas in developing countries around the world has been, and still is, fuel wood. Not only is wood used for cooking and heating, but it is also used by many small industries, including brick and cement making, and cottage industries such as refining palm sugar and small-scale distilleries. The consumption of wood for fuel in most developing countries is four or five times greater than processed wood, including paper. In many countries such as Ethiopia and Burma (now Myanmar) large areas of forest were cleared for fuel, both for domestic use and for making the bricks needed for building houses, temples and other public buildings. The Emperors of Ethiopia had to move their capital several times because they had denuded the area of trees. As we shall see later, the attractions of Addis Ababa as the location for the capital were such that it forced the Emperor to find a solution to the shortages of fuelwood rather than upping sticks and moving again.

In Myanmar, formerly know by the British as Burma, the Irrawaddy river flows more or less south through the whole length of the country, but after passing Mandalay it makes a turn to flow west and then southwest for about a hundred miles (one hundred and sixty kilometres) before turning south again near the city of temples at Pagan.

From Mandalay southward the river valley is very wide and flat, and being in the rain shadow of the Arakan mountains that lie between it and the Bay of Bengal is also rather arid. In the early 1980s the people living in the area began to struggle to find sufficient wood as most of the trees had been already cut, much of them in the distant past for making the bricks to build all the temples.

Figure 17: The temple city of Pagan, central Burma

Figure 18: Felling the last tree near Pagan, Burma

The government was keen to borrow funds from international donors to fund a large tree-planting programme. Since government officials had no experience in preparing such projects they called upon FAO to organise a training course in project preparation using the fuelwood project as the case study for the exercise. There were about ten trainees from the government, of whom one was from the secret police to keep an eye on what was going on. Unfortunately, one of the instructors sent by FAO, who had worked in the country before and knew the Director of a local research station, rashly decided to take half a day off from the approved itinerary in order to visit his friend. it seems that this was reported back to the capital, which was still in Rangoon (Yangon) in those days, and the following day the team leader received an order that the instructor must return to Rangoon immediately.

The training exercise involved visiting households to find out how many used wood for fuel, how much they used, where they got it from and how much they paid for it. The results were clear; shortages were causing serious hardship to many households that were having to collect from further and further afield, or pay a rapidly increasing price. We also looked at the availability of land that could be used for tree growing, and found that many farmers were willing to plant up part of their land with trees, if they could get suitable seedlings and good advice. Based on the information gathered we were able to help the Burmese trainees to formulate and present a project proposal to try to attract the necessary funding.

Figure 19: Old lady blowing to boost the fire to heat the pot, near Pagan, Burma

Figure 20: A household's stock of firewood, near Pagan, central Burma

Sahel and Drought

In response to the severe effects of recurrent droughts in the Sahel, that began in the early 1970s the United Nations established the United Nations Sudano-Sahelian Office that became widely known by its acronym, UNSO. It later transformed into, first the Office to Combat

Desertification and Drought and later it became UNDP's Drylands Development Centre. Despite its efforts the droughts persisted and the deserts continued to expand so that by the early 1980s starvation was widespread through much of the region. Ethiopia was particularly badly hit in 1983 and the suffering led among other things to 'live aid' concerts to raise money to help feed the starving.

In the forestry sector this stimulated interest in tree species that were drought resistant and it also created a fashion for 'multi-purpose trees' that could not only provide timber, but more important fuelwood as well as fodder for livestock, edible beans, nuts or fruit, shade and shelter and even nectar for honeybees. Many donors began to fund projects for afforestation and forester's enthusiasm for an Acacia tree (*Acacia nilotica)* grew as it was found to have all the properties that people needed. It would grow in a wide range of conditions, and had very deep roots, which not only helped it to withstand drought, but also meant that it did not compete strongly for the limited supply of water in the topsoil with food crops that could be grown between the trees. For a period in the 1980s it was considered as the 'miracle tree' that would solve the problems, especially in drought stricken Africa. The 1990s proved a bit wetter, and droughts became less frequent and severe, and so the caravan moved on to other things.

Establishment of ICRAF

As mentioned earlier, when describing some of the work undertaken to try to improve the management of tropical forests, in 1976 the global financial crisis meant that the United Nations Development Programme (UNDP), which had been funding a lot of the work, was suddenly starved of cash and cancelled all further projects. Since the UN FAO relied heavily on implementing UNDP projects for much of it's income, it began to switch its attention to desertification and rural energy, which were becoming the new 'flavours of the month'.

These topics required a completely different approach and were more about supporting farmers, diversifying livelihoods and improving productivity of marginal lands, particularly in areas prone to drought. The energy crisis, which had developed in parallel due to the sudden and sharp rise in the price of oil had the effect of highlighting the fact that most of the rural population in developing countries relied mainly on wood as their primary source of fuel. Thus the development agencies started to seek the magical cure for all these problems and discovered agroforestry. In fact, foresters had long known about the idea since we were taught about *'the Taungya system'* at university in the mid 1950s. This was a traditional land management system used in Burma (now Myanmar) whereby the British foresters who were then in charge, (one of whom was a lecturer), allowed local farmers to cultivate the ground between young teak trees that had been planted in forest where most of the older trees had been felled. The young trees are planted at quite wide spacing, partly to minimise cost and partly to avoid the need to do a thinning of rather small trees after a few years, and so the ground between the trees could be used to grow a variety of food crops.

In Indonesia, the Dutch had established Teak plantations in Java in the 1890s and they used a similar technique, which there was called *'tumpangsari'*. This approach was extremely beneficial both to the farmers and the foresters, because the farmers had the extra land to increase their output, without having to convert the forest, and the forester got their trees weeded and fertilised for free. With the plantations to be grown on a rotation basis, the farmers would just move to the next years planting area, though they usually could grow crops for a couple of years before the trees became too dense. This system was still being practised in Myanmar in the 1980s as seen during a visit there at that time, and in Java until about the year 2000. After that it's a different story that we will come back to later.

All this interest in agroforestry highlighted the fact that it needed both foresters and agronomists to work out the best practice under different situations. The Forestry department of FAO were among the leading exponents of agroforestry and the then Director-General, Dr. Ken King, persuaded a number of donors to establish a new specialised agency to undertake research and investment in agroforestry. In 1978, the International Centre for Research in Agroforestry (ICRAF) was formally established in Nairobi.

Ethiopia and Addis Ababa's 'Green Belt'.

Ethiopia is a fascinating country, with the western half lying mainly on a mountainous plateau to the west of the Great Rift Valley at elevations up to about 4,000 metres. The plateau is deeply dissected by a number of rivers, including the Blue Nile that flows out of Lake Tana in the north west of the country and makes a huge arc south-east then south then southwest and finally westwards into Sudan to link up with the White Nile. Standing on the edge of the Rift Valley the ground drops away sharply, but not precipitously, for about 3,000 metres, so that it is almost impossible to see the ground below and all the upper slopes are covered in forest. The Rift Valley extends northwards to become the Red sea and the low lying eastern part of Ethiopia in the valley bottom is hot and arid.

It seems that in the past as mentioned earlier, Ethiopians tended to cluster around their emperor's capital until all the resources had been depleted, when the whole population would up sticks and move to a new location. Such moves happened many times until they arrived in the place that is now Addis Ababa, which turned out to be so pleasant that the Emperor did not want to move again, despite the fact that the stocks of firewood in the surrounding area were heavily depleted. The story goes that the Emperor was discussing the problem with a French railway engineer, who was prospecting to build a railway line from Djibouti to Addis. Since at that time most of the railways in Africa used wood as the fuel for producing steam, and some of the railway companies were planting trees along their tracks to provide fuel for the future, the engineer advised the Emperor to have large areas planted up with *Eucalyptus* trees as a solution to the problem. The Emperor thought this a good idea and by good luck or through good advice chose to plant a species called *Eucalyptus globulus,* which is a mountain species in Australia and it turned out to be well suited to the conditions around Addis.

As a result the Addis Ababa Green Belt came into existence, and though somewhat degenerated by the early 1980s was then still in existence. The city lies towards the southern end of the main plateau as the land begins to slope down towards the Awash Valley. The green belt was established on the slopes above the city in the early days of the twentieth century and the trees have been cut regularly ever since. Over the eighty or so years since the Green Belt was established, the population of Addis had grown substantially, and the quantities of wood cut daily were consequently growing. Every day hundreds of old women could be seen with huge bundles of firewood on their backs carrying them down to the city. When one of the trees of this species is cut, it very quickly sprouts many new shoots from the stump, referred to as coppicing, by foresters, and the coppice shoots are much smaller than the original tree, which makes them much easier for elderly women to cut. However, after the coppice shoots are cut the next lot of shoots are much weaker as the stump gets older and begins to rot, so that eventually the trees must be replaced. This was not happening, and so the production of firewood from the Green Belt was steadily declining, while the demand was steadily rising; a recipe for disaster quite soon.

Fortunately for the Ethiopians, the World Bank came riding to the rescue with a large project to rehabilitate and extend the Green Belt and establish another Green Belt around a town on the shores of Lake Tana. The Danish and Finnish governments also joined the party with support for plantations around two other towns, one to the south called Nazret beside a lake that resembled the Sea of Galilee and the other called Debre Birhan to the north overlooking the Rift Valley. At this time, Ethiopia was beginning to recover from one of the longest droughts, which had caused widespread starvation. The immediate threat had been overcome with the provision of large quantities of food aid and the people were glad of paid labour planting trees until their next season crops could be ready. The results of the tree planting were generally good and within a few years the supply of fuelwood increased. In the plantations around Debre Birhan there were some problems with mole rats damaging the trees, but the villagers gradually learnt how to deal with them.

Figure 21. Addis Ababa 'Green Belt'' with young coppice in the foreground and degraded plantations behind (1985)

Travels on the Amazon

In the late 1970s travel in Amazonia was still difficult. One boarded a Boeing 737 in Rio de Janeiro or Brasilia for a 'milk run' around the major urban centres in the north east and north of the country, so that by the time we reached Belem there had already been 3 or four landings and take-offs. The landings always seemed to be hard, I don't know whether because the pilots were taught to make sure that the plane got onto the ground and stayed there, or because the runways were short and the approach too fast, but it seemed to be the same regardless of the weather. Belem and Manaus are attractive cities in the old town centres but Santarem, situated where the river Tapajos joins the Amazon was more chaotic. It lies at the end of a spur of the Trans-Amazon highway, which was under construction at that time. About sixty kilometres south of Santarem there are extensive rubber plantations established in the 1920s by the Goodyear tyre company. The trees suffer badly from leaf cutting ants that continually defoliate them and reduce yields. No solution has yet been found, and so rubber grown in Asia is much more productive, despite the fact that the tree species is a native in Amazonia. Further south on the same road is the Tapajos National Forest, where FAO is supporting the development of a management plan to demonstrate sustainability.

As part of the investigations we go to visit a number of trial plots established by EMBRAPA (The Brazilian Agricultural Research Corporation) in the forests around Santarem. Some of the plots are to the west down the Amazon river and as there are three foreigners and three Brazilians, we split into two groups. One group will go by speedboat and return by road, and the other group will go by road and return by speedboat. Travelling by speedboat is interesting seeing the vegetation along the riverbanks close up, but one doesn't get a real feel for the enormity of the river, because there are many islands, and for most of the way we are travelling in relatively narrow channels between islands and the shore. The plots are interesting and give us some useful data, and then it is time to return to Santarem. The group whose turn it is to use the speedboat includes an American Professor, who is a large gentleman. When he boards the speedboat there is little freeboard left. The engine is fired up and the boat moves off wallowing low in the water. It should get up enough speed to aquaplane more or less on the surface to reduce drag, but clearly the motor is struggling. As the boat slowly disappears from sight round a bend in the river, we see the boatman frantically trying to reposition the professor to get a better weight distribution, but the engine noise seems to indicate that he is having little luck. The noise fades into the distance and our group board the jeep and return to Santarem arriving well before the speedboat party who had a very slow journey.

The main topic of interest during this visit was the Brazil nut. It was the first time I had seen the complete nut, as usually by the time they get to supermarkets in the UK it is only the segments from inside the shell. The tree grows well and produces abundant nuts rather bigger than a cricket or hockey ball, but just as hard and heavy. Once the tree start to bear fruit the income for farmers is good, but it takes time, and the practice of agroforestry, where the farmers grow food crops between the trees provides income or sustenance until the trees are ready.

Riverboats in Bangladesh

The coastline at the head of the Bay of Bengal, which comprises the delta of the Ganges and the Brahmaputra rivers, was originally covered in mangrove forest. The western part of the shore in western Bangladesh and stretching into India is still covered in mangroves; an area known as the Sundarbans, and still the home of tigers. The mangroves on the eastern half of the coast of Bangladesh have largely been cleared and so the low-lying land is at the mercy of typhoons coming up the Bay of Bengal. When these coincide with high tide and high water in the rivers, enormous areas are flooded. Because of the huge silt load in the rivers the channels are constantly changing and new islands appear all the time. During the dry season, the islands are exposed and are quickly covered in grass, which offers tempting grazing for the local cattle, so there is a race between the farmers and the Forest Department, who want to plant mangroves to stabilise the land and protect it from erosion. The principle town in the delta is Khulna, which we reached by road after a daylong drive from Dhaka. My main impression of the journey was the traffic on the road, not motor vehicles, but ox-drawn carts, rickshaws and people on bikes and walking. In the morning for a while most of the traffic was going in the same direction as us, until we came to a village, after which most of the traffic was coming towards us. As the day progressed the flows seemed to be reversed. The land is intensively cultivated, with rice paddies in the low lying areas and crops of jute on the slightly elevated and drier land. The jute is now all processed in Bangladesh and India, but was originally established to supply the factories in Dundee, Scotland, where the technology for refining and processing the jute was developed.

The Forest Department have inherited a number of riverboats from the British built in the second half of the nineteenth century and still going strong. They were specially designed for the river with a broad beam and a shallow draft so that they could navigate all the channels among the many small islands. Inside they are fitted out with very comfortable sleeping quarters, galley and main cabin for eating and relaxing. The prospect of a couple of weeks on the boat seemed almost a holiday. The boat cast off from Khulna and cruised through a series of meandering river channels in a generally southeasterly direction. The following days were spent systematically visiting and measuring all the mangrove plantations and trial plots to try to identify the factors contributing to success and failure. In the evenings the boat moored at various villages and the crew would go ashore to purchase supplies. After about ten days we finally reached the town of Barisal.

The holy month of Ramadan started about the same time as the trip began, and so everyone was fasting during the daylight hours. That year, the monsoon, which should have been under way had not yet started and so food was in very short supply. In fact our diet for almost the whole two weeks seemed to consist of eggs, fried, boiled, poached, scrambled and in omelettes, sometimes with a little rice, and usually followed by pineapple for dessert. By the time the trip was over I did not want to see another egg, and in fact have hardly eaten one since.

It was clear from our inspections that given careful attention to planting and management, and protection from villagers who wanted the land for grazing the mangroves would grow extremely well and within a few years would form a natural barrier against the winds and tides, as well as stabilising the new land. In fact it followed what would be nature's way had there been no interference from humans. This illustrates the dilemma that is particularly strong for foresters where human populations have expanded so much that their short-term needs take precedence over the longer-term conservation of a habitable environment.

As we travelled around the delta we could see tall structures, usually just inland from villages at the river's edge. It turned out that these were refuges for the villagers during storms so that they could shelter well above the level of the floods. We were told that they had been constructed with World Bank funding a few years previously, but most were now too dangerous to use. It seemed that the contractors who had been charged with constructing the buildings had used second hand cement in order to reduce costs and increase their profits, and so the concrete was now crumbling. So that's what happens to so much aid money!

Troubles for the Tobacco Industry

The tobacco industry had been under attack for many years for contributing to lung cancer, but in the mid 1980s it came under attack from a totally different quarter. Anti-smoking activists began to claim that the tobacco industry was contributing to tropical deforestation, because tobacco farmers in tropical countries were cutting down forest to provide fuel for curing (drying) their tobacco. Amongst the claims being bandied about was the statement that "for every three hundred cigarettes one tree has been cut down". While no evidence was produced to support such a claim, it also clearly suffered from the lack of any definition as to what size of tree is involved. Assuming that farmers would cut fairly small trees, a tree might weigh around eighty to a hundred kilograms, while three hundred cigarettes would contain about three hundred and ninety grams of tobacco, and this implies that each kilogram of tobacco requires about two hundred and thirty kilograms of wood to cure it, which seems an excessively high figure in the light of the amount of energy contained in that amount of wood.

The tobacco industry had no firm data of its own, since most of the wood is cut and collected by individual small farmers, who do not record or report how much wood they use. Some farmers have their own tree plantations, but many just collect it free from nearby forest. The trade Association that represented the tobacco industry therefore decided that an independent assessment of the actual wood consumption was necessary and would enable the industry either to respond to its critics with real data or, should the critics prove right to promote greater efficiency and promote tree planting amongst its farmers. The Association requested our consulting company to submit a proposal for carrying out an assessment, which in due course was accepted. They decided that seven countries, out of sixty nine tropical countries producing tobacco, should be sampled as these seven accounted for about two-thirds of all flue-cured tobacco in tropical countries and provide a good cross section of tobacco growing conditions across the tropics. They also include smallholder production where each farmer may have just one curing barn, to large-scale production by large farmers that may have ten or more barns. The countries chosen were Argentina and Brazil in South America, Kenya, Malawi and Zimbabwe in Africa and India and Thailand in Asia.

Figure 22. Tobacco ready for loading into curing barns in northern Thailand

Local assistants were recruited in each country and a standard methodology developed for measuring the wood being consumed for the flue curing of tobacco. Each of the assistants in turn recruited a small team and all were trained in the method. The idea was to select twenty to thirty farmers who grew and cured their own tobacco in each country at random from the local tobacco company's list of farmers. Flue curing is done in barns, usually about four metres square and five metres high, which have wooden poles laid across them from wall to wall at slightly less than one metre intervals horizontally and in layers at similar vertical distance, on which the freshly harvested tobacco is hung in bundles. Underneath the barn is a furnace, and the flue from this furnace passes through the barn in such a way that the heat from the flue is well distributed throughout the barn.

Each farmer was asked to set aside slightly more than enough wood for curing one barn-full of tobacco, and this stack of wood was carefully measured to get both the stacked volume and the solid volume of the wood in it. Samples, in the form of discs were cut from a selection of the logs and these were measured for weight and volume, and then dried in an oven to drive off all the moisture and get the moisture content and the dry weight and volume. (Volume was measured using Archimedes' principle, whereby the sample was immersed in water and the weight of the water displaced provides a measure of the volume and therefore the density of the wood). This information enabled the total dry weight of the stack of wood to be estimated and hence the amount of energy that it contained. After the curing was finished, any wood left over was measured so that the amount actually used for the curing process was established, and at the same time the cured tobacco was weighed as it was removed from the barn. A total of two hundred and thirty three curing cycles were measured during the study across the seven countries and for each cycle the amount of wood required to cure the measured quantity of tobacco was therefore known.

As might be expected the results showed a lot of variation between farmers, but the farmer using the largest amount of wood among all the farmers studied used only about twenty six kilograms of wood per kilogram of tobacco or about one tenth of the amount implied by the anti-tobacco activists mentioned above with their claim of one tree for three hundred cigarettes, assuming just a small tree. The average varied from country to country, with Malawi having the highest of almost thirteen kilograms of wood per kilogram of tobacco and Argentina the lowest at just less than five kilograms per kilogram. This made it clear that the claims by the anti-smoking lobby were highly exaggerated and showed that the tobacco industry worldwide accounted for slightly less than one per cent of all wood fuel consumption.

The study showed clearly that far less wood is needed if it is well dried beforehand, and further gains could be achieved by ensuring that the barn was as full as possible of tobacco. The other interesting finding was the lack of understanding of the combustion process by many of the farmers, which was a major contributory factor to their high fuel consumption. The furnaces were all equipped with doors, with small vents, so that once the fire had reached the correct temperature the door could be closed and the air supply controlled to keep the fire going at the optimum temperature. However, many of the farmers said they preferred to us freshly cut wood, because it burnt more slowly, while well-dried wood burnt away very rapidly. The latter is undoubtedly true if the wood is supplied with plenty of air as a result of leaving the furnace door open, and the farmers had not appreciated that if the door was shut and the supply of air restricted, dry wood would burn much more slowly and hotter and hence longer. In fact, it was only necessary to keep the furnace door open because the wet wood needed much more air in order to get the temperature high enough to drive off the moisture. The evaporation of the water was in fact constantly cooling the combustion process, like continually putting some water on the fire, but many of the farmers did not appreciate this

The study also showed that farmers could be energy self sufficient if they had about the same area of land with fast growing trees as they had for tobacco. However, many of the farmers with relatively small plots needed all their land for tobacco in order to make a living and so could not spare land for growing trees, and thus relied on collecting wood for free from any nearby forests. Although tobacco farmers require more wood fuel than other farmers, in most tropical countries wood is still the major domestic fuel in rural areas for cooking and heating water for washing and bathing, and many farmers are faced with the same dilemma in not having enough land to grow both their food and their energy crops and so they still look to the forest for the latter. It was interesting, when travelling in Cambodia some twenty years later in the tobacco growing region, to see roadsigns provided by one of the tobacco companies advertising the farmers' tree planting scheme to achieve self sufficiency in fuel.

Although the main purpose of the study was to investigate the use of wood for the flue curing process, it became clear that a lot of other wood was used in the process. Some tobacco is sun-cured, rather than flue cured and the bundles of tobacco leaves are hung to cure in the sun on wooden poles supported by a wooden framework, and this accounts for a considerable additional quantity of wood. Added to that is the wood needed to make the cigarette papers and filters, packets, cartons and the pallets used in transporting the cartons. Apart from showing that there is considerable scope for improving the efficiency in the use of wood-fuel, it also showed that more wood than tobacco goes into a cigarette, especially if one includes the packaging, the cigarette paper and the filter, all of which are made from wood.

Wood for Whisky Making

During the 1980s the high oil prices stimulated interest in biomass fuels, even in the UK. The Department of Energy at the time had a special Energy Technology Support Unit (ETSU) that was promoting many forms of alternative energy by providing grants and other support for innovative commercial scale trials. One such trial was the conversion of the boiler at a major whisky distillery in Speyside, Scotland to utilise wood residues for fuel. An entrepreneur had negotiated with the distillery owners to install a new boiler that could burn wood residues and to guarantee supplies of suitable residues from the forests around the north east of Scotland. One condition of the financial support from ETSU was that the whole process should be monitored closely for a year in order to assess the real technical and financial feasibility of the concept. The boiler was duly installed and instrumented to measure the fuel input and energy output and the working temperatures achieved, since the fuel was likely to be relatively wet, as it would mostly come from the branch-wood and tops left over after trees were felled and harvested.

The entrepreneur leased a chipping machine mounted on a four wheel drive articulated skidder with a grapple crane that could pick up all the branch-wood and feed it into the chipper which in turn blew the chips into a large hopper mounted on the back of the skidder (see figure 23) and trials were carried out to compare the time and energy used for harvesting different types of residues organised in different ways. In some trials the branch-wood and tops were just left wherever they were cut, in others the residues were pushed into windrows by the tree felling crews so that the chipper could move systematically up and down the rows and in others the tree fellers moved all the residues into large piles so that the chipper did not have to move around while chipping. The windrowing and piling of the residues added to the costs for the tree fellers, but reduced the costs for the chipping, and it was important to know which gave the overall lowest costs.

Figure 23. Skidder mounted chipper with grapple crane used for harvesting tree residues as boiler fuel for a whisky distillery

It took the chipper about half an hour to fill the hopper, when it would move to the nearest roadside and discharge the chips into the back of one of the large trucks that were parked and waiting. Three trucks operated a shuttle service between the forest and the distillery and each could carry about three hoppers full from the chipper, so trucks were departing about every two hours.

In order to assess the efficiency of the operation and get accurate comparisons between the different ways of organising the operations the engines that powered the chipper and the skidder were fitted with fuel flow gauges that enabled the amount of fuel being used over very short time periods to be monitored. From this information a comprehensive picture could be built up of the actual effective time used for chipping the residues as well as the amount of energy used effectively for chipping and wasted in idling and for just moving round the site.

In addition to the chipping operations, the haulage contractor agreed to make the tachograph records for the trucks available for analysis so that the time, travel speed and efficiency of the haulage could be measured. This turned out to be a most informative exercise as with a little practice it became possible to tell from the record exactly where the truck was at any particular time. For example, about six miles from the distillery there was a hairpin bend in the road on a steep hill and the trucks had to slow right down to negotiate it safely. It very quickly became possible to identify patterns on the graph that corresponded with junctions, bends in the road and built up areas. From this we were able to plot the movement of the trucks on maps, all in the days before GPS became commonplace, and measure the travel time on different classes of roads. The results were very clear that travel was far faster on A class roads than on B roads, which in turn were faster than C roads. This enabled the delivery cost to be calculated quite accurately for any chosen block of forest, according to the distance that the truck would travel on each class of road from forest to distillery.

All this information allowed a very detailed schedule of the costs of the wood residues delivered to the distillery with different species, harvesting regimes and source locations. The results showed that the overall cost of the energy for the distillery was just competitive with fuel oil, but great care had to be exercised to keep the costs competitive. Some forests were just too far away to be worth harvesting, or even though close enough for crows involved tortuous journeys, because of the mountainous terrain and so were too expensive. A small fall in the price of fuel oil could have a big impact on the overall area of forests that could be economically harvested. These restrictions mean that to be economic in future, supplies of biomass will need to be dedicated for a particular user. This really means special plantations established within a limited radius from the user, rather than using whatever residues happen to be available.

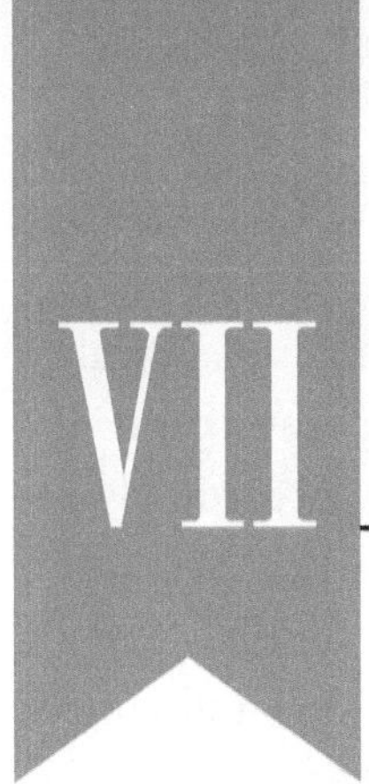# VII The Rise of Intermediate Technology: Small and Medium Enterprises

t was in 1973 that E.F. Schumacher published his book *Small is beautiful* and started the Intermediate Technology Development Group to promote appropriate technology in development. The book challenged the conventional economic dogma of the time that for growth, large-scale operations and consumption were essential and emphasised the need for sustainability and the use of technology that was of a level of sophistication appropriate to both the human capacity that will use the technology and the resources that the technology would be used to process. In forestry this could mean the use of animals such as oxen, horses or elephants for extracting timber rather than tractors that used large quantities of fuel and also had to achieve high outputs in order to cover their costs. For many foresters this argument is very convincing, since it has been the use of large machines that has contributed most to the destruction of much of the tropical forest. Had animals been used for extraction the supply of logs would have been much smaller and therefore sustainable, while the costs would be higher making logs more expensive and thus promoting high efficiency in the use of the limited supply.

A Tornado in Tonga

The plane from Nandi in Fiji to Tonga was delayed, without any reason given as to why, but an hour or so later things became clear, because news came through that a severe tornado had just struck the island. It became doubtful if it was worth making the visit, which was to look at possibilities for setting up joint ventures with European wood manufacturing companies in some of the Pacific islands, including, Fiji, Tonga, Samoa and Vanuatu. However, news came that an Australian military plane with emergency supplies had landed in the island and damage, though widespread was not totally devastating. Accordingly the scheduled flight left about 36 hours late and touched down at Nuku'alofa in Tonga safely an hour or so later. As we came in low over the main island we could see damaged buildings and coconut palm trees lying in tangled heaps all over the place, but there seemed to have been very few casualties. It is not a long drive to the hotel, which is situated on the waterfront at the northern edge of the town. It had taken the brunt of the winds and waves and had sustained substantial damage, but still had some rooms that were habitable, and there wasn't exactly a crowd of visitors in need of accommodation, only myself and one other person.

The main island of Tonga does not have any forest, only palm trees, but about twenty kilometres to the southeast is the small volcanic island of 'Eua which has some forests on the mountain slopes with a limited potential for harvesting. Alternatively, some entrepreneurs in Tonga had proposed importing logs from Papua New Guinea or Vanuatu, which had relatively abundant supplies, and processing the wood in Tonga for export to the USA or Europe. It was an interesting idea, inspired by Singapore, which has no timber resources of its own, yet has significant exports of wood products. On the island of Tonga, which has an abundant supply of coconut palms, there was a modest factory, sawing and processing coconut palm trunks into roofing tiles and flooring. The products are quite marketable, but the wood has a high silica content and is very hard on saw blades, which require special tungsten tips that are costly to import. There was no shortage of enthusiasm among both government officials and entrepreneurs to expand wood processing on the island, the problem is to find European companies willing to invest and transfer the technology and help with marketing the products.

On departure, a senior official from the ministry escorted me to the airport and arranged for me to use the VIP lounge. The plane was delayed, and so I was glad of the facilities and enjoyed some refreshments and passed the time reading. As the plane approached, I thought it time to visit the toilet before departure, and on entering was surprised by both the size of the facility and the lavishness of the fittings, so that it looked more like a throne, with polished brass handrails. I think I would have disappeared down the hole, had I sat on it. As I emerged, the look on my face must have conveyed my astonishment to the attendant, who informed me that it had been built for use by the King, Taufa'ahau Tupou IV, well known as a very large gentleman. He died in September 2006 at the age of eighty eight.

Small and Medium Industries in Honduras

The USA provides a lot of aid to Honduras and one of the minor topics of interest is forestry and the processing of wood. The country is interesting from a forestry point of view, because it has natural pine forests in the mountains and tropical lowland forest that is rich in mahogany, amongst other species along the coast in the southwest corner of the country. The native pine species is called the Caribbean Pine (*Pinus caribbea*) and it has been found to grow well in many tropical countries in plantations. The remaining dwindling resource therefore has a special value as a gene pool for an important tree species. The pine forest that is being harvested is mainly sawn into squares (called cants) and these are exported to the USA and Canada, which adds very little value in Honduras and employs very few people. The government was keen to do more processing in the country both for domestic consumption as well as for export. But with little expertise or capital the prospects of establishing a large processing complex without foreign capital and assistance were pretty remote. In fact a Canadian company had invested in 1977, the year before my visit, in a sawmill to produce sawn pine lumber for export to Canada, but it was very large and used the latest technology and so was mainly operated by Canadians and only employed local people for very menial tasks like sweeping the floors.

At that time the Oxford University Forestry Institute was working on the design of a low cost solar heated drying kiln for timber and in Scotland we were running a small company producing charcoal using both mobile steel kilns that could be moved to the wood in the forest and larger fixed kilns that used mainly off-cuts from sawmills delivered to the kiln site. There were also a growing number of designs of small portable sawmills becoming available, some using chain saws with cutting guides, others using circular saw blades and still others using horizontal band-saws, The concept to be investigated in Honduras was to bring these three types of equipment together to create a small-scale integrated wood processing facility. It could produce kiln dried sawn timber suitable for house-building and furniture making and almost all the residues could be converted to charcoal for heating and cooking. An analysis of the likely financial performance of such a facility indicated that it would be profitable and would need very little capital investment compared with the large-scale sawmill, designed in Canada that had just been installed in the country.

One lesson from this, which applied also to the charcoal company in Scotland, is that it seems to be possible to operate at a small scale producing mainly for the local market or at a very large scale for national or international markets, where economies of scale are essential in order to compete, but not at an intermediate scale which produces more than the local market can absorb. This also has implications for developing a business, and it is clearly very difficult to make the transition from a small-scale cottage industry to larger scale manufacturing, where marketing becomes an essential aspect of management.

Figure 24: Polythene solar heated timber drying kiln set up in Honduras

Forest Extraction in Brazil

Brazil's expanding manufacturing sector needed steel in large quantities, and with very limited coal resources, charcoal became the preferred source of carbon for making steel (see Figure 13 and the section on charcoal in the previous chapter). Initially the wood needed to make the charcoal came from clearing areas of the dryland *Cerrado* forest, but a combination of economics and environmental concerns pushed the companies into establishing plantations. In the state of Minas Gerais quite large areas were established on quite steep hilly land that presented some problems for harvesting since the type of heavy machinery used in the flatter areas of natural forest with big trees was quite unsuitable

The solution to this problem was to develop a simple but elegant cable system. A continuous cable was laid out in a rectangle around the area to be harvested and passed through a pulley wheel driven by a small motor located at the top of the active vertical leg. At the bottom of the active leg it passed through a second pulley mounted on a small mast about two metres tall. At each of the corners was a horizontally mounted pulley wheel anchored in the ground that turned the cable through ninety degrees to make the rectangle. The driven pulley had a simple device for tensioning the cable so that the driven pulley gripped it all the time regardless of any load on the cable. On the active leg of the cable circuit the direction of movement was downhill and chokers looped around the butt end of logs were attached to the moving cable so that the log was dragged downhill, and the chokers were released just before they passed over the pulley on the tower.

This system is a good example of appropriate technology since it was cheap to build and operate and so did not have to achieve high outputs in order to minimise overall costs. Having been manufactured locally the skills and spare parts needed to maintain it were readily available and also cheap. The whole operation was in fact much lower cost than more sophisticated systems that might have been imported from Europe

Figure 25: Simple cableway system used in Brazil for harvesting plantation grown Eucalyptus logs

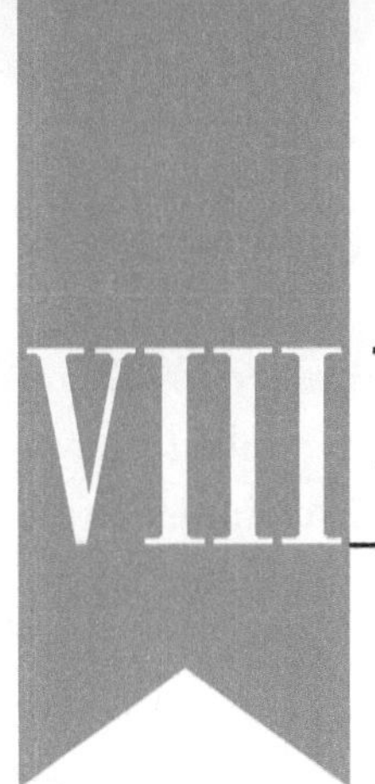

VIII — Trying to Tackle the Demand Side 1983-90

The International Tropical Timber Agreement

The increasing concerns about tropical deforestation led to the negotiation and eventual signing of an International Tropical Timber Agreement in 1983 between a number of major countries producing tropical timbers (the producers) and some of the major industrialised countries that imported tropical timbers (the consumers). It was hoped that this would lead to a more orderly trade in tropical timber products and help to promote sustainable management of the tropical forests, since logging was held to be largely to blame for their destruction.

One of the outcomes of this Agreement was the establishment of the International Tropical Timber Organisation (ITTO) to act as an executive office and implement the agreement. ITTO organised regular meetings of the members and initiated a wide range of studies into all aspects of tropical forest management and exploitation. One of these studies that became known as 'the Incentives study' investigated ways and means of providing incentives to companies engaged in harvesting, processing and trading forest products and managing forest resources to carry out their activities in a sustainable manner. It was implicit in the study, that managing forest resources sustainably would be more costly than merely exploiting the resource and that, private sector operators would therefore require some kind of financial incentive to encourage them to take a longer term perspective. The big question therefore was – would it be possible to provide incentives in such a way as to achieve the desired sustainable resource management, without distorting the market and creating perverse consequences?

In order to determine whether financial incentives are required, and if so how large they should be and how they might be applied, it was first necessary to better understand how value is added to timber along the chain from standing tree to a piece of furniture in someone's home. Looking at about thirty countries heavily engaged in trade in tropical timber products, either as producer or as consumer countries, it is possible to recognise three broad categories of product: Logs and cants (which are just squared logs), primary products (sawn timber/plywood) and secondary products (joinery/furniture/utensils etc.) The logs are produced in the tropical producer countries, but the production of primary and secondary products may be done in the producer country, an intermediate country or in the consumer country. Each stage in the process involves the use of resources of labour, energy, capital and raw materials each of which has a cost, and the company will add value if the sales value exceeds the total of these costs.

As a contribution to the study, I constructed a model of the value added chain from the forest through harvesting, primary and secondary processing with three variants; the first had the whole chain in the producer country with exports of finished products to consumer countries; the second had harvesting and primary production in the producer country, exporting semi-finished goods to the consumer countries which undertook the final processing; and the third had just the harvesting in the producer country with log exports to consumer countries, which undertook the further processing to finished product. The data used in the study was collected from the producer and consumer countries visited which were averaged for each stage in the chain. Thus labour, energy and other costs used in the producer country and the consumer country were an average value of those found in producer and consumer countries respectively. There were generally differences in these costs between producer and consumer country but much less within either of the country groups.

The study had a number of interesting results that were somewhat counter-intuitive, but which were explained by the model. For example, undertaking all the processing in the producer countries resulted in significantly higher costs for the finished products than exporting logs and doing the processing in the consumer countries, despite the lower labour costs and raw material costs in the producer countries. The reason was that the much higher levels of labour and capital productivity and better utilisation of raw material in the consumer countries, more than offset the modest advantage of lower labour and raw material costs in the producer countries. The study also revealed that value of the logs used in most products only made up about ten per cent of the value of the final products, and that taxes, such as VAT, levied in consumer countries accounted for more of the final price than the wood. As a result, the rich consuming countries were benefiting from the trade in tropical wood products far more than the poor tropical developing countries whose forests were being cut down, even when much of the processing was done in the producer country.

More Value Added

In Papua New Guinea, the Australians had been running a forest service for some years after the country gained its independence in 1975, while training locals to take over in the future. Initially they had focused on identifying areas suitable for logging and then purchasing rights to the timber from the tribal owners. Having secured the rights to the timber they awarded concessions to companies to undertake the logging. After some years concerns began to be expressed that the country was not benefiting from the industry, because almost all the felled logs were being exported at very low prices and relatively little employment was being created locally. It was strongly felt in some quarters that more emphasis should be put on to processing the logs. The government responded by requiring companies to process the logs in the country and export semi-finished or finished products. Calculations suggested that this would increase revenue as well as employment.

For about eight years prior to 1979 there had been lengthy debates about the direction and content for a National Forest Policy, which culminated in the publication of a White Paper in 1979. Although the country had gained independence from Australia four years previously it was still largely run by Australians, though financial aid from the

Australian government was being reduced and one of the aims of forest policy was to increase government revenue to fill the expected funding shortfall. Compared with the extent of the country's forest resources, the forest industry was very modest and was mainly run by foreigners. The emphasis in the new forest policy was on increasing exports, but considerable concern was being expressed in many quarters, especially the Ministry of Finance, that much of the financial benefit from the industry was being lost because of what is known as transfer pricing. This is done by companies having one invoice for the export documentation to show to the taxman with a price that barely covered their costs, while the invoice delivered to the overseas customer was much higher, even though the customer normally paid the shipping costs, and was paid into an offshore bank account, usually in Hong Kong. The Institute of National Affairs, an independent research and policy 'think-tank' therefore decided to sponsor a study to examine various issues related to forest policy, which they asked me to undertake.

To find out what was going on, data and information was gleaned from many sources to put together a balance sheet for the industry as a whole showing the costs of all the factors of production for each of the segments of the industry such as sawmilling, plywood production and wood chips for pulp, and comparing the result with the total revenue from sales on the domestic and export markets. The results did indeed show that, while exporting logs was profitable, processing in the country was so inefficient, with so much wood being wasted, that the value of the exported products was less than would have been obtained from just exporting the logs. This may have been partly due to the transfer pricing which understated the value of the exports, and comparing the reported export prices with international prices for similar products strengthened this suspicion, but it is almost impossible to prove.

On the bright side, the performance of a cooperative called the Alliance Training Association was a shining example of how to promote entrepreneurship in developing countries. The Association ran a sawmill in a relatively remote area in the highlands, processing mainly the local southern Beech *(Nothofagus)* and they supported enterprising locals to run small businesses as contractors, particularly for transporting the logs. They provided a long-term contract to an individual, which enabled him to get a loan from a bank to purchase a truck, on condition that the truck was brought to the Association's factory site every week for service, maintenance and book-keeping. This ensured that the operator kept his truck in good condition and kept proper and detailed accounts. Over a few years at least one individual had been able to progress to buying a further truck, and when I last heard his enterprise had grown to become a major transport company.

Furniture for the Miners in Russia

As Russia began opening up under '*Perestroika*', led by Mikael Gorbachev, many state enterprises began to look to the west for investment and advice to improve their situation, since central planning was no longer being undertaken. Most wood processing enterprises had been established to meet local demand for forest products ranging from lumber and plywood through to paper and furniture. The town of Vorkuta, just north of the Arctic Circle and west of the Ural mountains, had grown up around a number of large coalmines. The first coalmine was opened in 1936 on the site of a reindeer farm and in 1943 the present town of Vorkuta was established. It was cold, even in August, and a biting wind was blowing as we stepped off an aeroplane after an hour's flight from Pekhora, the capital of the Autonomous Republic of Komi. We were whisked in a fleet of black cars to an office in the city centre, where we met with a group of officials. At the time, the miners were on strike, and one of their grievances was lack of furniture and fittings for their apartments.

Some years previously, when the mines were opened up, and apartment blocks built to house the miners and their families, a furniture factory had also been established to produce all the furniture and fitments needed, for kitchens, bedrooms, dining rooms and living rooms. Vorkuta is located near the northern end of a railway line that eventually links into the main rail network, and huge trains, a kilometre or more in length, pass down the line daily carrying the coal to wherever it is needed. When the trains return they bring supplies including logs from forests in Siberia and elsewhere. The furniture factory takes in the logs, and processes them into a range of furniture items. The enterprise showroom is full of very nice looking tables, chairs, beds, wardrobes, kitchen cabinets and so on, but the miners complain that they cannot get what they need and have to wait for so long before anything that they order, is ready.

Inside the factory, the problem becomes clearer, because every spare piece of flooring is covered with stacks of components and partly finished items. The workers manning the saws and lathes can hardly get to their machines because of the piles of unfinished items. It is difficult to understand why this should be so, but the explanation is that the hardware and the soft fabrics needed to finish everything, like hinges, glue, paint, locks and catches, door handles and foam and cover fabrics for chairs and sofas has not arrived, because some central planner forgot to put them in the quota for the factories that produce them. So much for central planning! The enterprise management were hoping that a foreign investor would inject some capital to enable them to buy the goods they need, from overseas if necessary. However, it is not just working capital they need. The equipment, like saws and planing machines is old fashioned and inefficient, and the factory is grossly over-manned, so that to get it to a standard where it could compete in the international market, would require not only investment, but also substantial redundancies. Added to this, the enormous distance by rail to a port from where products could be shipped, and the cost of operating for much of the year in sub-zero temperatures would make it difficult to sell the idea to an investor. It is difficult to explain such considerations to people who have been brought up in a centrally planned and controlled economy, and who think that someone just has to give the order.

Arrangements were made for us to return to Pekhora by road so that we could see some forest areas and also another wood processing complex. The visit to the forest, involved a two-hour drive from the main road, to reach one of the blocks where they were then harvesting. Periodically along the road we passed large trucks hauling logs. The road was in an appalling state with huge quagmires in places and no proper surface, so that the trucks were taking a real battering. No wonder the Russians go in for such rugged construction; it may survive the conditions, but it would surely have been cheaper to invest a bit more in decent roads and save fuel as well as wear and tear on the machinery. The forest reminded me very much of what I had seen in eastern Canada, with pine and spruce predominating and birch and aspen in patches scattered throughout, only they were the more familiar European species of Scots pine and Norway spruce. Clear felling was being applied using a harvester based on a military tank chassis. Stopping on the way back to visit areas felled a few years earlier, it seemed that the Russians had the same problem as the Canadians, in that the young saplings needed to regenerate the area were very scattered and quite sparse away from the edge of the uncut forest. Nothing was being done about replanting, and so it could take a very long time for these huge cleared swathes to become fully tree covered again.

Leaving the forest we travelled on to the 'Mega wood-processing complex' a real eye-opener. Several plumes of smoke and steam could be seen in the distance, and then the smell of sulphur grew stronger until finally this huge factory came into view. In the complex were two pulp-mills, a huge saw-mill, a plywood mill, a particleboard mill and a factory making animal feed from cellulose. A total of over seven million cubic metre of wood were consumed annually. The plywood mill had recently been installed by a Finnish company and was making a very nice product from birch, which is in increasing demand. The other mills were in a more dilapidated state. The railway from Vorkuta runs past the complex and beside it there was a large black stinking lake. This was where they had disposed of lignin residue from the plant that made animal feed by hydrolysing cellulose and adding yeast. Apparently over two million tonnes had been dumped before they got round to installing a boiler that could burn the stuff. The hosts asked me if I had any ideas about how this material might be used. In modern pulp mills this black liquor is burnt to generate power or steam for the pulping process, but it seemed that it was easier to get some coal from the mines in Vorkuta than to utilise the processes own residues. I suggested they might consider making charcoal out of some of the other solid residues from the sawmill and then soaking the charcoal in the liquor so that it could be burnt without having to change the furnaces. I don't know if this suggestion was ever followed up.

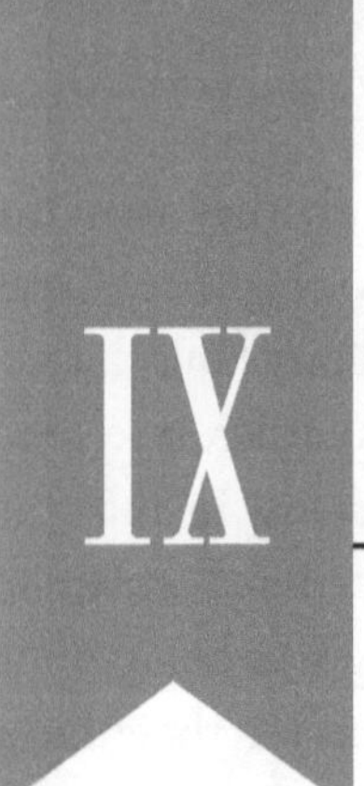

Sustainable Forest Management and the Environment (1992-97)

Forest Radio Systems in Indonesia: Can Better Communications Stop the Rot?

Indonesia is a vast country, with a land area almost the same size as the four mainland Scandinavian countries, the three Benelux countries, France and the UK combined. This comparison doesn't fully demonstrate the extent of the country, since Indonesia is comprised of three large islands and three quarters and a half of two other large islands and up to 13,000 medium and small islands, which are spread out over a maritime area about 5,000 kilometres from east to west and almost 1,000 kilometres from north to south. The total population at over 250 million is about twenty five per cent greater than the nine European countries compared for size, but almost sixty per cent of the Indonesian population live on Java island, the smallest of the large islands a little bigger than the UK, while the Indonesian part of the island of Borneo, called Kalimantan is almost as big as Finland and Sweden combined with a similar population.

Much of the so called 'outer islands' (that is outside Java) was densely forested until recently, but when Suharto assumed power in 1967 he rapidly arranged for large areas of forest to be offered as concessions to foreign and local companies to exploit. The sad story of what happened as a result has been described in a number of books and reports, but one contributory factor to the rape of the forests was the Ministry of Forestry's inability to monitor and control what was going on because it had no way of communicating rapidly with officials in remote areas. In the beginning, the forest department did not have the staff to properly supervise the companies, and by the time they had built up their staff capacity, many of the logging companies had become so powerful, and had such strong political influence, that the Ministry of Forestry could do little to control them, even had it wanted to. Bribery and corruption were often strong incentives to turn a blind eye.

Despite this somewhat gloomy picture, there were people in the forestry department who were concerned about what was going on and wished to be able to communicate rapidly between the field and the offices at province and headquarters level. They wanted to have better information about what was going on, and in particular to improve the response to calamities such as forest fires and floods, which seemed to be becoming more severe. The state forestry company, Perhutani, which had a monopoly over the forests in Java, which included valuable teak plantations had had a radio communication for some years, which had been very effective in the management of the forests.

The Ministry of Forestry therefore sought financial and technical assistance to carry out a feasibility study for an integrated forestry radio communication system for the whole of the rest of the country. The British government's Department of Trade agreed to fund such a study and also to provide some soft loan finance for installing a system should it prove economically viable. I was asked to lead a team of specialists that was assembled and after exhaustive studies, calculations and negotiations a final report was produced that demonstrated that the overall national value of the benefits from having a radio communication system through better management of the forest, better control over illegal logging and forest fires and improved contact with local communities would be far greater than the costs of building and maintaining it. However, the total cost for even a basic system that could be extended later, was about three times greater than the funds that had been committed by the UK. It was therefore proposed that the system should be installed in three phases, with the UK government funding the first phase. Later the EU was to agree to fund the second phase and the USA the third phase, so all seemed rosy.

The first phase was duly installed with equipment supplied by Phillips Radio Communications (formerly Pie Radio of the UK) and the third phase was completed relatively quickly with equipment supplied by Motorola (a US company). The EU sent consultants to review the proposals for the second phase, and eventually approved grant finance to build the system in the provinces planned for the second phase. More EU consultants arrived, and it became clear that a French telecoms company was interested in the contract to install the second phase, but they did not have the equipment capable of operating at the frequency selected for the system. The frequency had been selected with care to give the best compromise between ability to function in forest clearings and the range to minimise the distances between repeater stations, and had been allocated by the appropriate Indonesian authority. It seemed that in Europe, the frequency concerned had been allocated as a special NATO frequency and the French company did not make equipment to operate at that frequency. They therefore lobbied hard to have the specifications changed.

The consultants were dismissed and more consultants sent out to argue for a change in specification, but they eventually proposed a complete redesign of the system to use a combination of the new public system then being extended and satellite phones. However this took no account of the additional costs that would have been required to extend the backbone of the new system to enable most of the forestry offices to be connected, as they are generally located in the smaller and more remote towns. The proposal would have kept the installation costs within the budget, but it would also have opened up the system to abuse and potentially enormous running costs if staff used the system for personal calls as well as official ones. If this were to happen, as seemed very likely,it could have bankrupted the provincial forest departments. The radio system, while a bit more expensive to install has very low operating and maintenance costs once in place, as calls are essentially free, while satellite and public telephone calls are charged by the service provider. As it happened, the satellite that the EU consultants proposed to use, was never launched. Eventually the EU cancelled the grant funding, and the forest department is gradually completing the system with its own funds.

The European Commission in a Fix

About 1990 the European Commission decided that it needed to know how much it was spending on forestry around the world. Initially it suggested that there were about forty projects in different countries run by several different Directorate-Generals. In fact in those days there were five Directorate-Generals within the Commission that had an involvement in forestry in one way or another. DG I was responsible for foreign relations with the so called ALA countries (Asia and Latin America) and included assistance with forestry matters while DG VI dealt with forestry within the EU that included certain overseas territories such as French Guiana and DGVIII dealt with the ACP (Africa, Caribbean and Pacific) countries in the same way as DG I. Then there was DG XI that was responsible for the Environment and DGXII that dealt with Research and Development. It turned out that all five DGs were dabbling in forestry in one way or another, although there was only one forester on the staff at that time sitting in DG VIII.

Visits to each of the DGs secured huge numbers of documents relating to projects of one form or another ranging from multi-million ECU, (as it was in those days before the Euro was created), to very small grants to NGOs planting a few trees somewhere. It turned out that some countries operated, what was called a STABEX scheme for commodities such as coffee where prices fluctuated from year to year. The EU supported these schemes by putting money into the fund when prices were high so that it could be drawn upon to support farmers when prices dropped. It seems that the result was rather like the wine lakes and butter mountains that the Common Agriculture Policy created and the funds just kept on growing, because the price that triggered the payments in or out of the fund was set too low, and in good times large amounts were accumulated and not much was paid out in bad times. Someone had approved the use of some of the surplus funds for tree planting projects from four different STABEX funds, but it was not counted as forestry because the department concerned was not supposed to have any forestry activities.

After examining all the files it turned out that there were more than three hundred forestry related projects that had been approved and funded over the previous five years, and that the total expenditure was over ECU 300 million, but because it was being spent by so many different departments no-one had any idea as to what it was being spent on or whether it was achieving anything. Since then the Commission has been drastically reorganised, but I wonder if they are any better at keeping track on where the money is going.

Trying Again in Indonesia (1991-2000)

This part of the story is the continuation of the efforts to introduce the basics for sustainable forest management abandoned by UNDP in 1976. When inflation died down and the oil price returned to a level that it had been a few years earlier, interest in tropical forest management declined and the world and the politicians became more concerned with economic growth again so that the emphasis in forestry turned to forest industries and growth in demand for forest products, although droughts and famines, especially in the Sahel focused some attention on desertification. Managing the tropical forests was more or less ignored, and in countries like Indonesia, huge concessions were awarded to companies to log. Although there were conditions attached to the concession licences on how the forest should be managed, these were not enforced, and the concessionaires made little effort to protect their forests from encroachment, that usually followed when roads were opened.

A decade later the environment began to attract attention again so that the UN announced that they were organising an 'Earth summit' to be held in Rio de Janeiro in 1992. The issue of tropical deforestation came up again as a topic to address at the conference and one consequence was that the British government suddenly announced that it would invest £100 million in efforts to introduce sustainable management in tropical forests. Indonesia was a target country. Following a mission by a number of experts a major Programme was agreed between the Indonesian and British governments to cover forest policy, forest management, conservation, research and training and teams of consultants were dispatched to prepare projects in detail.

At that time the British aid programme was under the Overseas Development Administration (ODA), which for forestry matters, had a rather incestuous relationship with the Oxford Forestry Institute (OFI) (no longer in existence)and the Natural Resources Institute (NRI), based in Oxford and Maidstone respectively. The two institutes had contracts with ODA to provide technical expertise, and very little work was offered to independent consultants despite there being considerable talent available. Much of this talent is concentrated in Edinburgh, which is home to the Royal Botanical Gardens of Edinburgh (RBGE), the Edinburgh

University Department of Forestry and Natural Resources (DFNR), the Institute of Terrestrial Ecology (ITE)[2], the Forestry Commission (FC) and LTS International Ltd, a forestry consulting company, of which I was a founder and Director. These five institutions had come together to form the Edinburgh Centre for Tropical Forests (ECTF), and together had a formidable range of talents and expertise covering the whole spectrum of issues of concern regarding tropical forestry.

Having had the projects that were to form the Programme designed, ODA turned as usual to OFI and NRI to implement the projects, but after almost a year it became clear that they did not have the capacity to do so, and so ODA was forced to go out to tender. This gave ECTF the chance it had been waiting for and an ultimately successful bid was submitted. The project and the institution were tailor made for one another, because the range of tasks to be undertaken by the projects was well matched with the expertise available. I travelled to Indonesia in October 1991 to take over the ailing project and breath new life into it, and by early 1992 all the projects were fully staffed and well under way.

The Programme consisted of five projects covering Forest Management at policy and implementation levels, Research, Training and Conservation and the programme was referred to as the Indonesia-UK Tropical Forest Management Programme (ITFMP). The consultants who had designed the Forest Management component had only made provision for a couple of high level advisers in the Indonesian forest department HQ, and had not designed the approach to be adopted to forest management nor to how field level implementation would be conducted. As a result during my first meeting with the Indonesian Director-General of Forest Utilisation, I suggested that we should consider trying to establish a pilot forest management unit along the lines discussed with his predecessors in 1975. His response was that the government had been planning to do just that, and that he had signed a decree to that effect. The Indonesian term for a Production Forest Management Unit is KPHP and so the next eight years was spent trying to get five such units established and operating.

After Suharto seized power in 1967, it was essential to generate revenue quickly and the vast forests were an obvious source of cash. The

---

[2] The Institute of Terrestrial Ecology later changed its name to the Centre for Ecology and Hydrology (CEH)

military, influential individuals and companies were invited to prospect and apply for concessions and within a few years much of the accessible forest was allocated. The concessions boundaries were generally sketched on old and inaccurate maps and were often irregular shaped blocks that bore little relationship to the situation on the ground. Latecomers got all the odd bits and pieces that had not already been grabbed, and so a provincial Concession map looked like a strange jigsaw puzzle.

Most concessions were drawn based on having river access and the concessionaires established camps and log ponds on a major river. From there they built roads into the forest, extending the road each year as the logging proceeded. These roads, which were meant for bringing logs **out** of the forest, also provided easy access for migrants to get **in** to the forest and clear some to set up a homestead and small farm. The soils under the tropical forest are generally very poor and the forest mainly depends on re-cycling the nutrients that are built up slowly over the years by the deeper rooting trees. When the forest is cleared this stock of nutrients rapidly disappears and so the migrant farmers had to clear more forest, as the yields on their first fields, or '*ladangs*', declined. After a few years the forest around the logging camps became very moth-eaten and more and more settlers arrived. The settlers generally used fire to clear the land, and as the forest became more and more open the fires could easily spread into forest and huge areas were laid waste.

Bringing some order to this chaos is quite a challenge; especially since most of the forest blocks that would be big enough to manage for commercial timber production have from ten to twenty communities living within them or around the border. Some of these communities were composed of indigenous people who had lived in, and depended on, the forest for generations, although usually not in the same location as in the past communities had moved regularly, though not far. Other communities were made up of groups of migrants that had found their way, usually from various parts of overcrowded Java or Sulawesi and still others were called 'transmigrants' as they had been moved under a government sponsored resettlement scheme. As a first step in establishing permanent forest management units, a major effort went in to establishing the external boundaries, which would have to be really permanent if sustainable management is to be achieved. Although the old concessions were intended to follow some kind of harvesting rotation that allowed the newly cut areas to recover before being harvested again after thirty five years, this became impossible because most of the forest that should be harvested for a second time in years thirty six onwards, no longer existed as forest, having all been converted to some form of agriculture or increasingly to oil palm or rubber plantations.

A lot of work was put into surveying and negotiating boundaries with the concession holders and the local communities and carrying out detailed inventories of the whole area that was expected to become a management unit to provide a basis for the long-term sustainable harvesting plan. These surveys once again demonstrated clearly that if the forest was left undisturbed it would rapidly recover. One area in central Kalimantan that was inaccessible and had been harvested about twenty years previously had recovered to more or less the same stocking as the un-logged forest. In contrast another area in Sumatra that was easily accessible and subject to rampant illegal logging had degraded so much that it was already past the point where it could ever recover, and in fact in 2008 I heard that it no longer was forest, having all been felled and converted to oil palm. By the end of the eight years the five management units were more or less ready to be officially declared and for implementation of the management plans to begin. A detailed Manual on the establishment and management of KPHPs in the national language (Bahasa Indonesia) was distributed widely, and the Ministry of Forestry was reorganised with a new Planning Agency that had responsibility for rolling out KPHPs across the country.

However, projects must end, and ODA having now become DFID under Mr Blair's new labour Government needed to do something different and a new flavour was therefore order of the day called 'community empowerment' which resulted in the follow up project going in a totally different direction.

It was therefore encouraging to visit Indonesia again in 2008 and find that the Forest Department is now actively trying to establish KPHs for Production, Protection and Conservation everywhere and all the concepts developed during the piloting have been adopted. In 2011, Indonesia was one of eight countries selected to receive funds from the Forest Investment Program (FIP), one of a family of funds for supporting activities relating to Climate Change Mitigation and Adaptation. These funds are managed and implemented by the World Bank in association with the relevant Regional Development Banks, which for Indonesia is the Asian Development Bank (ADB). The Forest Department made it clear that this funding would be used to support the establishment of a number of KPHPs in different parts of the country. It is very gratifying to know that the nine years of hard work is bearing some fruit.

Oil Palm Boom

While efforts to introduce sustainable forest management in a number of pilot areas were going on, the price of Crude Palm Oil (CPO) began to rise as demand overtook supply. Many Malaysian companies that had extensive plantations in their own country, turned to Indonesia to find cheaper land to increase their crop areas. When CPO prices are high even relatively low yielding plantations on marginal sites can look financially attractive, and as a result huge areas of forest in Sumatra were cleared and converted to oil palm, much of it on soils quite unsuited to long-term sustainable production. However, looking at price data for CPO over many years, it was apparent that the long-term average price was around US$350 per tonne, but during 1992 and 1993 it rose sharply to over US$500 per tonne, and this made oil palm plantations look extremely attractive financially. Needless to say, within a few years the Asian Financial Crisis occurred and the price of CPO fell back to its more normal level for a few years, so that some plantations were then abandoned. All the major commodities such as rubber, oil palm, coffee and others are subject to cyclical movements in prices, partly influenced by the general global economic situation, but partly by temporary imbalances between supply and demand. As demand overtakes supply, the price tends to rise, and this stimulates an expansion in the crop areas. After a few years the new production comes on stream and supply overtakes demand so that the price falls again. It seems that it is always easier to expand the area of the crop, when the prices rise, rather than increasing the productivity of the existing crops, and so the forest is under continuous pressure to be converted to provide more land for these various commodities.

At the same time as the boom in oil palm, some of the Indonesian companies that had grown rich from exploiting the natural forests for logs to make plywood were deciding to branch out into the manufacture of pulp and paper. With the Chinese, Indian and other Asian economies growing rapidly, the demand for paper products in the Far East was expected to increase rapidly in the coming years.

Three huge pulpmills were under construction in Sumatra, and although they were ostensibly intended to use plantation grown wood as raw material, plantations require a huge investment and take time to grow. The pulpmills were therefore very glad to get hold of the timber cut from all the forest being cleared for oil palm plantations, and of course selling the logs to the pulp companies more or less covered the costs of clearing the land for the oil palm. This made the oil palm plantations even more profitable.

The Ministry of Forestry had been through a process some years previously by which a proportion of the more accessible forest on flatter ground was classed as 'Conversion Forest', which meant that it was available for clearance and conversion to other land uses. About a quarter of the forest area was so classed. The Forest Law defined 'forest' as being land with more than twenty per cent of the area covered by tree crowns. Any area with less than this proportion of tree cover was considered as 'degraded' forest and would thus be eligible for clearance and conversion to plantations or some other use. This provided a huge incentive for unscrupulous logging companies to fell as many trees as possible in order to reduce the stock to below the twenty per cent level. They could then get permission to clear the rest.

Figure 26. Clearance and burning of forest for oil palm

Shortly after the logging boom began, the Ministry Forestry introduced a levy on logs cut that had to be paid into a 'reforestation fund'. The initial idea was that the levy should be a kind of bond, and would be refunded when the Ministry of Forestry was satisfied that the planting of trees in gaps left in the forest after logging had been done properly. However, the amount that had to be deposited was far less than the cost of doing any planting, and so it was easier for the companies to pay the levy and neglect the planting. As a result the reforestation fund began to grow into a substantial sum, and later became a kind of slush fund that could be used for many things, including the development of a civil airliner. An early idea for the reforestation fund was to use it to promote tree plantations, and the Forest Department announced the "Industrial Forest Plantation" scheme (known by it's Indonesian acronym HTI). Under this scheme, commercial companies could get low interest loans for plantation projects, or could establish joint ventures with state timber companies, who would contribute their share from money provided by the fund. Thus a combination of oil palm and pulpwood plantations led to the clearance of most of the remaining forest on the huge lowlands on the eastern half of the island of Sumatra. The Ministry of Forestry claimed that the forest being given to the oil palm companies was classified as 'Conversion forest' and the land being used for industrial tree plantations was already badly degraded and therefore qualified for financial support under HTI. In fact the boundaries of the different forest classes were being continuously re-drawn. Areas of Production forest became 'Conversion forest' allowing them to be cleared and areas, generally on higher and steeper land, that had been set aside as Protection forest became Production forest to maintain the overall area of production forest. Meanwhile Illegal logging was ensuring that the forest was being steadily degraded.

Illegal Logging and 'Certification' in Indonesia

The term 'Illegal logging' is used here to refer to any logging that is not officially sanctioned as part of a management plan or special licence and therefore is neither recorded nor subject to payments of royalties or taxes. It can take many forms and be carried out by many different parties. Some is done by companies that have a licence to harvest in a concession, but cut more trees than their permit or cut trees from adjoining areas that are not part of the management plan. This often happens when a concession adjoins a national park or nature reserve and the company cuts trees within the protected area. It can also entail cutting trees from within blocks that have already been harvested before the time scheduled for further harvesting, either by the company having the concession or by others. These others may be from local communities or they may be from far away, but they are usually sponsored by a *'Cukong'* which is the name given to an entrepreneur that organises the operation. The *Cukong* may be operating for his own profit or may be working on behalf of powerful business or political interests that often own an unregistered sawmill. The cost of obtaining a permit to harvest logs officially is very high, because of the need to bribe many officials, and so it is generally cheaper just to go ahead and log without a permit.

Because such logging is unrecorded it is impossible to know with any certainty just how much timber is harvested in this way, but the visible deterioration of the forest and many stories told by concession companies, local communities and even government officials suggest that it is widespread and involves huge volumes of logs. It is therefore important to try to make an estimate of the scale of the problem, if anything is to be done about controlling it and promoting the supply of 'Certified' timber. Three approaches were adopted in efforts to put a figure on the likely volume of logs harvested illegally in Indonesia. The first step was to determine the capacity to process logs by all the different factories. It was known that there are many small sawmills operating without a license or with one issued locally that was not registered with the Ministry of Forestry. The latter tended only to have records covering the larger integrated mills that were processing wood primarily for export. Surveys were conducted in a number of areas by volunteers who travelled *incognito* and used casual and informal techniques to locate all the sawmills and obtain an estimate of the quantity of logs each processed in a year. Having obtained an estimate of the total capacity to process logs and the amount processed annually from both official sources and the additional surveys it became possible to construct a balance of total recorded supply of logs from all official sources and compare this with the consumption of wood by the processing industry. This suggested that the actual annual consumption of logs was about double the recorded volume, and so it was assumed that the volume of logs harvested illegally was about the same as the 'official' harvest.

Another way to arrive at an estimate of the illegal harvest was to estimate the amount of wood used within Indonesia for house-building, furniture, paper, handicrafts and other uses and compare that with the production from all the factories after deducting exports and adding in imports of wood products. This is less accurate than the first method but actually gave a very similar picture, and it became apparent that almost all the recorded production from the big licensed wood processing factories was being exported while the domestic consumption was mainly coming from the small unlicensed sawmills. The government had been so preoccupied with exports and generating foreign exchange that it had ignored the need for wood by its own population. The third approach used involved looking at the area that was identified as having been logged over from satellite observations, which was part of a national forest inventory. Many areas that had been logged were no longer forest and so historical records of forest cover had to be used to arrive at a figure for the total area logged, which could be compared with the area reported to still be primary forest. Applying the average volume of logs harvested to the total area harvested gave an estimate of the total volume harvested over the 30 years since serious logging began. When this was compared with the recorded volume harvested, it again indicated that the volume of logs that had gone from the forest was more than double the volume that had been recorded as harvested over the thirty-year period. Thus all three methods led to the same conclusion that the volumes of logs being harvested illegally was in fact very large and was both costing the government dearly in terms of lost revenue from royalties and taxes and was degrading the forest so that it would not be able to sustain such levels of harvesting in the future.

Using the model described in Chapter VIII above, with the specific information on costs and prices being collected for all the operations and products involved for our economic work, we could analyse the financial performance of the sector, and I was then able to combine it with all this information on the supply and demand for timber to construct a detailed model of the whole sector covering both material and financial flows. This not only revealed the scale of the financial losses that the government was incurring as a result of the illegal logging, but more interestingly it suggested that the developing pulp and paper industry was not generating sufficient surplus to cover the cost of servicing and repaying its debt. At the time when we reached this conclusion, no one in government, or industry or among donors seemed very interested, but a year later when the Asian financial crisis hit it became strikingly clear that the model predictions had been correct.

Later, this issue of illegal logging emerged in a different context, when scientists concerned with climate change began to estimate the quantity of carbon dioxide emitted from the burning and decomposition of woody material in forests around the world. The widespread existence of this illegal logging has serious implications for consumers of wood products who wish to have only wood that is 'certified' as having come from sustainably managed forests. A company having a forest concession may have all the necessary paperwork to show that their title to the forest is legal, and they may have a management plan that ensures a balance between the volume of logs harvested and the re-growth of the forest after harvesting, but a buyer of wood products has no way of knowing if such documents are valid and are being complied with by the concession operator.

The concession agreements that most companies hold are not titles to the land, only permits to harvest logs from within a specified area. Frequently these concessions overlap land that has been the traditional territory of long-time local inhabitants. Management plans may look fine on paper, but do not guarantee that the concessionaire is only cutting the number of trees approved from within the area specified for harvest each year. A number of companies have set themselves up as 'certifiers' to check all these details independently and issue certificates that comply with the Forest Stewardship Council's standards. Would that it were as simple and as straightforward as that. The certifying companies are paid by the concession company, which creates a risk of conflicts of interest. They are also under pressure from both the timber trade and from environmental activists to certify as much forest as possible to increase the availability of 'certified' wood and to demonstrate the success of the campaigners in tackling illegal logging. Certainly in the early days many concessions were given certificates, despite the fact that they did not comply with the first requirement of having legal title to the land. In Europe and North America, where a company or individual has legal ownership title this is no problem, but in most tropical countries there is no such thing as legal title, to most land. In Indonesia all land that has not been titled, which is almost all, outside Java, is State land, but people have occupied substantial areas, frequently since before Indonesian Independence. Large areas are designated as 'forest land' but this only means that it comes under the jurisdiction of the Forest Department, who do not have any form of title to it

The problem was well illustrated in Java where the majority of the commercial forests are teak plantations established by the Dutch in the late nineteenth century. A state company, *Perhutani* has a monopoly over the management of these forests, and was one of the first tropical timber growers to apply for 'certification'. Following inspections and checks they were given a certificate in 1995. Word soon got around that teak wood from the *Perhutani* forests was certified, and very quickly, large numbers of small enterprises, many run by foreigners, appeared in Java and Bali who claimed to be producing furniture from 'Certified' timber. However, there was no chain of custody process in place that could distinguish between wood that had been felled as part of a management plan, and wood that had been felled illegally. The Asian financial crisis, know locally as

'Krismon' (Indonesian acronym for Monetary Crisis) hit around this time, putting large numbers of people out of work, many of whom found that felling some teak trees and selling them to these small furniture factories was a good way to make some money for survival. Many teak forests that I visited in the 1980s and 1990s no longer exist, and when the certificate expired, it was not renewed.

Illegal Logging in Laos

Laos is a small landlocked country, about ten per cent bigger than the United Kingdom with a population of only around 6.8 million (slightly more than Scotland). It will figure again in other topics. The northern part of the country is vey mountainous, while the southern half has an extensive plain in the west along the Mekong river, and a mountain chain known as the Annamites forming the eastern border with Vietnam. The country was reported as being densely forested by early French travellers in the nineteenth century, and the government claims that sixty to seventy per cent of the land area was forested as recently as around 1950. The forest cover is now reported to be down to about forty per cent of the land area, but only about three per cent is relatively undisturbed dense primary forest. This represents a loss of almost five million hectares of forest or about five hectares for every family in the country. The remaining thirty seven per cent of the land area has tree cover, but generally very disturbed and degraded. How could such a small population of people destroy so much forest in such a short space of time?

Part of the answer is warfare, as Laos was heavily bombed by the Americans during the Vietnam war to try to stop the supply of armaments from the North to the Vietcong in the south, which were being transported along the 'Ho Chi Minh Trail' that passed through Laos. Figure 27 later shows the extent of the bombing in one small part of the country. However that does not explain how so much forest has disappeared. The Government, with the help initially of Sweden carried out a series of national forest inventories every ten years starting in 1982, which showed that forest cover has declined from about fifty per cent at the first inventory to forty two per cent in 2010. However, this does not show the decline in the quality of the forest, which has become steadily degraded so that the estimated volume of the trees has declined by about a hundred and forty seven million cubic metres over the twenty eight years from 1982 to 2010 or a reduction of about five million cubic metres annually. This compares with the volume of logs harvested over the same period, as reported by the government of around eleven million cubic metres; so the next question is: where have the one hundred and thirty six million cubic metres gone?

Some has gone up in smoke and has contributed to the increase of carbon dioxide in the atmosphere, but much of it has almost certainly gone to the country's neighbours of Vietnam, China and Thailand. All three countries have large populations and have depleted their forest resources over the years so much that they have introduced logging bans. Various analyses of timber supply and demand in Vietnam, one that I made while working there on another projects, and one by Meyfroidt and Lambin (2009) showed that the wood industry in Vietnam was using about ten million cubic metres more logs every year than were harvested in the country according to government records. Some of this was imported from Africa and South America, though trying to get import statistics out of the government agency was worse than getting *blood out of a stone*. Much of it came from Laos and Cambodia, but there is only anecdotal evidence to prove it.

On one occasion, while travelling in Vietnam, we stayed overnight at a small hotel on a main road leading from Laos to the town of Dong Ha in central Vietnam, just north of Danang, and I was kept awake after dark by the noise of heavy truck rumbling past. As my room was at the front, overlooking the road I was able to count forty nine of them and could see clearly that they were all carrying squared baulks of timber, which must have come from Laos, as there is no forest between the hotel and the border. In the morning we passed one of the trucks broken down at the roadside, and we estimated that each truck was carrying about 100 cubic metres of timber. (see figure 27). After reaching Dong Ha we travelled south towards Danang and passed a number of sawmills with large logs, that could not possibly have come from anywhere in Vietnam, which has very little primary forest left from which such large logs could have come. We dropped into one of the sawmills and asked the owner where she got her logs from, to which she immediately replied "from Laos". She then showed us an invoice for the logs from a Vietnamese construction company in Laos from whom she said he had purchased the logs

Figure 27. Lorry carrying squared logs from Laos, broken down at the roadside in Vietnam.

On another occasion I travelled with a Lao colleague to the border post near Ban Pakha on the most southerly road from Laos to Vietnam in the Lao Province of Attapeu. The road to the border passes through a large tract of forest that has been designated as a National Conservation Area, and in the middle of it was an enormous clearing, of maybe a hundred hectares almost filled with stacks of logs. Later on the way up to the border we passed a number of huge heavily laden trucks carrying logs. At the border we decided to have some lunch at one of the small stalls beside the road, and it turned out that it was run by the wife of one of the customs officials, who came for his lunch while we were there. As there were several trucks laden with logs parked nearby, we managed to engage the official in casual conversation, and eventually learned that during the dry season forty to fifty such trucks passed across the border each day, despite the fact that there is a ban on log exports in place in Laos. The official informed us that the trucks had special papers signed by a Minister authorising the trucks to cross into Vietnam. He said that they were reparations to Russia for aid provided during the war, but was not clear as to whether they were reparations from Laos or from Vietnam. A likely story!

Two years later I was asked to make a study of the contracts awarded for, what is termed 'salvage logging' or the harvesting of trees from areas that have to be cleared for some form of infrastructure development such as a hydro-power dam. During visits to the field, I managed to obtain copies of a number of such contracts, and one of them explained the flow of logs witnessed at the Ban Pakha crossing. A Vietnamese company had been awarded a contract to harvest the logs from the area that was expected to be flooded by a large dam under construction by another Vietnamese company. The contract provided for a 'loan' to be paid into an account in Laos that would be repaid with the value of the logs harvested, in return for all the necessary permits to take the logs to Vietnam. The 'loan' was for US$ 130 million, and after six years of harvesting the nominal value of the logs exported was only US$ 60 million according to the Lao government's official log price, though the real value of the logs at international prices was probably more than enough to repay the loan.

In the same province of Laos, and further west on the same road there is a huge rubber plantation extending to over 10,000 hectares established by yet another Vietnamese company called Huang An Gai Lai. (HAGL) The plantation had been established in forest that had been designated as Production forest, and was included in a World Bank funded project that had developed management plans that would ensure that the area was managed 'sustainably' to provide logs for the local industry. However, HAGL had built a village for the athletes attending the Southeast Asian Games held in Vientiane (the capital of Laos) a few years earlier, and was paid with the logs from the forest cleared for the rubber. Kenney-Lazar (2010) provides a more detailed account of the deal. Although the awarding of the concession and approval to harvest the logs was approved at Ministerial level, and could therefore be considered as 'legal' the approval is contrary to the Forest law, which forbids clearance of forest designated for Production, Protection or Conservation.

The company is now constructing an airport near its offices and factory in Attapeu and one wonders how the Lao government will pay for it?

Forests and Floods

With much of the human population living in river valleys and especially in river deltas, flooding has always been a hazard for many people, but over the past couple of decades they seem to be more widespread and devastating. In east and southern Asia there have been serious floods in Vietnam, Cambodia, Thailand, the Philippines, China and Indonesia, and these have generally been blamed on loss of forest cover, so that most of these countries have banned logging and clearance of forests in an effort to reduce the problem in the future.

The Annamite Mountain range, forms the watershed between the Mekong river to the west and the various rivers that cascade down the mountains on the eastern side into the South China Sea, but from about 17° latitude southwards the mountain divide is within Vietnam leaving a large part of the country known as the Central Highlands, in the Mekong river basin with rivers flowing westwards. The three major rivers that rise in the Central Highlands are the Sekong, Se San and the Srepok and the first passes through southern Laos, while the other two flow down just through Cambodia to merge just before they disgorge into the Mekong at Stung Treng. The forests in the headwaters of these rivers in Vietnam were badly damaged during the Vietnam war, often through the use of chemical sprays by the Americans, but also as a result of felling trees for reconstruction after the war. Down river, Cambodia is used to the seasonal expansion and contraction of the huge inland lake, Tonle Sap, but in recent years the pattern of river flow seems to have changed so that the changes in water depth are becoming more extreme and this has been blamed on the loss of forest cover in the river basin.

The question was asked; "is it possible to demonstrate that the changes in river flow and increased flooding are a direct result of the loss of forest cover?" A difficult question to answer, as it proved, but an interesting one to investigate nevertheless. A friend of mine who is a Geographical Information System (GIS) specialist and I were asked to investigate. In the 1970s, while working in Bangkok, I remembered an occasional visit to the Mekong River Committee office and wondered what had happened to all their records. At some stage the MRC had been disbanded, and later a new body, the Mekong River Commission established with new headquarters in Vientiane. What luck then, when browsing through the library at the Asian Institute of Technology in Bangkok to find racks filled with the Mekong Hydrological Journal published by the old MRC covering much of the period from the mid 1960s to the mid 1990s that included records from a number of river gauging stations and rainfall stations in the central Highlands. Unfortunately many of the records were irregular, with some years missing and sometimes recording seems to have ceased altogether. It was possible to get records for a ten-year period from 1984-1995 for six river gauging stations and five rainfall stations on the two rivers. The data was in tables in these huge leather bound journal volumes, but fortunately a student happy to supplement her stipend was willing to scan the records needed and provided us with excel spread sheets, which made the analysis manageable; so much for modern technology!

The results were not absolutely conclusive due to the limited data sets but they did give a strong indication that in the decade between 1984 and 1995 the run-off in the river increased quite substantially for a given amount of rainfall and this coincided with a rapid and substantial loss of forest cover, especially from the steeper slopes. It may not have been a coincidence that the only time that the level of the Mekong river at Stung Treng exceeded the flood danger level was in 1994, when it rose enough to flood almost half of Cambodia. Analysis of the rainfall and river flow data did show quite clearly that the river flow levels were only really affected by very major rainfall events associated with tropical typhoons, which dumped more than 100 mm rain in a twenty four our period. The effect of such downpours was different at different times during the season. If one occurred early in the rainy season the effect was relatively modest, while one late on in the rainy season caused a rapid and substantial rise in the river level; this is probably because early in the rainy season the soil is dry and can absorb large amounts of water, but as the season progresses the soil becomes increasingly saturated and additional water runs off rapidly into the rivers. It is therefore clear that forests do have some beneficial effect but maybe more on soil erosion than on rainfall run-off. However, it is known that forests transpire more water than grassland or annual crops, and therefore tend to dry out soils more and to greater depths. It might therefore be reasonable to assume that run-off would be less and slower from under forest than from other land uses.

Meanwhile my friend was collecting and analysing old satellite images of the whole Lower Mekong river basin going back to 1973. He was able to find four complete sets covering the whole of the lower basin for four periods, 1973, 1985, 1993 and 1997. The area of the part of the river basin that we were interested in was 62.5 million hectares, of which 34.5 million hectares or about fifty five per cent was forest covered in 1973. By 1997 the area of forest had been reduced to only 16.7 million hectares, a staggering loss of almost 16 million hectares in just twenty four years. About half the lost forest had been on steep slopes where soil erosion becomes a serious matter, so that the removal of the forest cover would almost certainly have resulted in accelerated run-off and hence would have increased the risk of flooding. This rate of forest cover loss is consistent with the picture drawn earlier regarding illegal logging in Laos.

A year or two later there were devastating floods in the southern Philippines on the Island of Mindanao and the City of Cagayan d'Oro, on the north coast, was particularly badly hit. By a strange coincidence I was asked to help with the design of an Integrated Watershed Management Project in the Philippines, which included one of the river basins in the Province of Bukidnon on Mindanao. The province is dominated by Mount Kitanglad, near its centre, and the river that flooded Cagayan d'Oro rises on the mountain along with four other major rivers. Two tributaries of the Pelangi river, which flows northwards from the mountain are the Muleta and the Manupali and these were the subject of a detailed study a few years prior to my visit.

The University of Auburn in the USA had assisted the local university with monitoring water quality over a couple of seasons in four sub catchments of the Manupali and had designed very simple low cost methods for monitoring river flow and depth and suspended sediments. The river flow was measured by suspending a float from a bridge on a fairly long string so that as the flow rate increases and the water level rises, the angle of the string from the vertical, changes. They were able to calibrate the angles against the flow rate in the river using a flow meter and so that local community representatives only had to record the angle of the string, which they did at regular intervals. The water depth was measured with a calibrated pole in the riverbed and water samples were collected for analysis at the University to measure the suspended sediments. The four sub-catchments had different degrees of forest cover and were subject to differing farming practices. The results after the two years of monitoring showed that the peak run-off after storms and the quantity of suspended sediments was lowest in the most densely forested sub-catchment and *vice versa*. For more details see Deutsch *et. al.* (1998).

The region generally is famous for the large fruit farms belonging to large corporations which grow mainly pineapples and bananas. Over the years they have been expanding their land holdings by buying out indebted farmers. The farmers then move up the mountain and clear more forest from the Mount Kitanglad Range Natural Reserve, which is meant to be a Protected area. The continued loss of forest can be expected to exacerbate the flooding and erosion problems in the various rivers. The Pelangi river has a very large hydro-electric scheme down river and the operators are already experiencing problems with siltation of the dam and excess wear on the turbine bladed.

Smallholder Plantations in Laos

Laos, which prefers to be referred to as Lao People's Democratic Republic (Lao PDR) is a mountainous and landlocked country, mainly on the east bank of the Mekong river. At the time of its colonisation by the French during the nineteenth century it was densely forested and sparsely populated. Most of the population was concentrated along the Mekong river, but there were many small communities of various ethnic groups that shunned the malaria infested lowlands and preferred to live in the mountains, where they practised shifting cultivation. Over time, the land used for agriculture in the lowlands expanded and the combination of spreading agriculture in the lowlands, shifting cultivation in the uplands and wars, resulted in the gradual depletion of the forests. By the early 1990s neighbouring Thailand had already exhausted most of its forests and in response to increasingly severe floods, imposed a logging ban in an attempt to protect the remaining areas. As a result Thailand became a net importer of wood and Laos was an obvious place to look.

At about the same time, the economy in Vietnam was beginning to take off after the American War and this stimulated an increase in demand for wood, accentuated by a sizeable furniture industry that grew up in the former Saigon, now known as Ho Chi Minh City, and developed export markets in Europe and North America. The forests of Vietnam were in a similar condition to those in Thailand, partly because large areas had been damaged or destroyed during the war with the Americans in the 1960s, and also because there had been uncontrolled logging by state timber companies to meet the demands of reconstruction. Meanwhile, China's economy was also growing and demand began to outstrip supply from their own forests. Many of their forests were also seriously depleted which was considered to be a major contributory factor to a series of severe floods on the Yangtze river, that killed many people and damaged much property. As a result, a logging ban was imposed in 1998 in China on most of the forests in the western part of the country. As mentioned earlier in Suriname, China was already importing logs from many countries, but the logging ban increased the need, and Laos and Myanmar (Burma) also became targets. In 1998, the year of the logging ban in China, the 'recorded' exports of logs from Laos to China suddenly increased fiftyfold and since then have matched those to Thailand. There is of course no record of the logs exported illegally, but analysis of the production and export of manufactured wood products in Thailand, Vietnam and China compared with the recorded log production in those countries suggest that the volumes of illegally imported logs must be enormous, mainly from Laos and Cambodia.

In 1978, China embarked on a program of economic reform and modernisation, which was followed in 1986 by a similar move in Vietnam referred to as *Doi Moi*, and a decade later Laos followed suit and also began reforming and opening its economy and looking for foreign investment. Hydro-power and minerals were added to logs as the major sources of foreign exchange. As mentioned above the official records of log production and exports suggested that the quantities of logs being harvested were quite modest, but the reality was very different with large quantities of illegal logs being shipped over the borders to the wood hungry neighbours. Eventually the government acknowledged that its forest resources were seriously depleted and so in 1997 agreed to a project to be funded by the Asian Development Bank to support the establishment of plantations, mainly by farmers to diversify their source of income. On the face of it, the project concept seemed a good idea, as the low population density meant that land was available, and being a poor country incomes and wages were low. With labour costs accounting for about three quarters of the cost of plantations, even if the farmers hired labour to do the work the plantations should have been profitable for the farmers.

The funds from the ADB were channelled through an Agricultural Bank and farmers could apply for loans to cover the cost of the plantations or to provide them with some funds while they worked on their plantations. Some of the money was used to buy seedlings and fertiliser and the rest was used either to pay labour or to provide the farmer with some cash for his own labour. What could possibly go wrong? The first problem was that the Agriculture Bank was set up to support agriculture which has annual crops, and the farmer only needs credit for a few months between planting and harvesting, while with trees the farmer gets no revenue for seven years or more, but the Bank still required them to pay interest on their loans each year, so the farmers became more and more indebted. The second problem was that the farmers were meant to have had advice from the national agriculture and forestry extension service to ensure that they established their plantations properly and looked after them; especially applying fertiliser and keeping the trees free of weeds. However, because it was individual farmers who chose to borrow money and establish plantations, each plantation was generally small, just one or two hectares, and they were scattered far and wide with just one or two plantations in a village. This made it extremely difficult and expensive for the extension agents to visit each farmer and check on progress. Records got lost and staff changed and eventually most of the farmers were left to get on, on their own.

The third problem became apparent after a few years, when the first plantations reached a size when they could be harvested. The wood using industry in Laos is small and fragmented and mainly geared to using large logs from the natural forests. Only one or two companies set themselves up to utilise the plantation trees and they required relatively large logs for sawing and could not handle the majority of the logs that were quite small. With the farmer's plantations scattered far and wide, most farmers had no ready market for their produce. Across the border in the town of Khon Kaen in Thailand there is a pulpmill that was having difficulty in getting enough raw material from local plantations, and so several Thai traders started to buy up logs from the Lao farmers, and shipping the logs to the mill in Thailand. Of course the traders paid very little to the Lao farmers, most of whom finished up in debt, while the traders made large profits, some of which was used to bribe Lao and Thai officials to smooth the journey to the mill. The Agricultural Bank almost collapsed because of the large volume on non-performing loans to the farmers for tree planting.

Having learnt the lessons from this first project we designed a follow up project in a radically different way. A new agency was proposed, which would organise farmers into groups, so that intensive technical support could be provided, and the funds were treated as equity to be shared between the agency and the farmer so that the risks are shared. The agency would also take a strong role in future marketing to ensure that the growers got a fair price, and it had an incentive to do so, because it recovered some of its equity investment when the crop was harvested, which could be re-invested in new plantations. The Asian Development Bank offered the government a substantial grant to set up this new agency and invest in smallholder plantations, as well as overseeing commercial companies that wanted to invest in plantations, to assist them in finding suitable land, and ensure that they followed the rules in the way they treated the local communities and the environment. The Grant was eventually approved, after much discussion, by the ADB Board.

The Lao Forestry Department appointed a senior staff member as Managing Director Designate for the Agency, to be called the Lao Plantation Authority, and I was asked to act as Interim Chief Technical Adviser. We established an office and started the process of recruiting staff and prepared an Operations Manual while waiting for the Cabinet to approve a Decree formally establishing the Authority. One day the Lao MD was called to a Cabinet meeting and set off in high spirits, expecting to get the green light. To his surprise he was asked if the ADB was likely to agree to put the funds into a 'Policy Bank' that the ministry of Finance had just set up. Not having an answer he left the meeting and returned to the office with a long face and asked me what I thought ADB would think of the idea. For reasons unknown, but probably political, the government decided at the last moment not to go ahead with the project.

There are many large multi-national companies based in Japan, Indonesia, India and Europe operating pulp and paper mills in the region, and supply of raw material is becoming an increasing problem for them. The word soon got around that Laos might be a suitable place to establish plantations, because of the very low population density and generally good growing conditions. In fact by about 2003, it was not just pulp companies that were looking for land in Laos, as companies came from many countries looking for land for rubber, sugar, cassava, maize, *Jatropha* for biodiesel and many other crops, so that pressure for land suddenly became severe. Most of the companies chose to seek land in the relatively flat plains on the east bank of the Mekong in the southern part of the country. The only problem was the fact that most of the area had formerly been a dry forest, rather than a rain forest, and the soils are both shallow and compacted. For tree growing the better soils are to be found further east and at higher elevations, nearer the border with Vietnam, where the soils are deeper, more fertile, and rainfall is higher, but these areas are littered with the UXOs (unexploded ordinance) as we shall see a little later.

Participation and Budget Support (1997-2002)

Stakeholders Become Involved

By the mid 1990s interest in sustainable forest management and the environment was waning and the new buzzwords were 'stakeholders' and 'participation'. While this was to be welcomed as an effort to ensure that communities settled in or near forests could became involved in decisions relating to the management and utilisation of the forest, achieving it in practice is far more difficult than the theorists working in development agencies and NGOs would have us believe. Many communities in developing countries are not homogenous and may consist of more than one ethnic group that can include, so called indigenous minorities and incomers from the more dominant ethnic group. In most Asian countries, there tend to be a very large number of indigenous minorities, while the population of each is quite small. Some such communities have been more or less in the same place for many generations, while others have either moved of their own choice or been moved as a result of war and other calamities relatively recently so that claims to traditional rights to land and forest are almost impossible to determine. By the year 2000 some of the negative aspects of 'participation' were becoming apparent, and are discussed in the book by Cooke and Kothari, *Participation: the new tyranny.*

In most tropical countries, there is a strong negative correlation between the extent of forest cover and population density. Where human population density is low there is still relatively extensive forest cover and communities tend to be small with just a few hundred families. Where population density is higher, forest cover is much less, communities are larger and the pressure on the forest, especially for logging illegally, is also much higher. In the past, private or state owned companies have been given licences or concessions to harvest logs with little or no regard for the rights and interests of the people living in or near the forest, who may have been very dependent on the forest for almost all their essentials bar staple food crops, but including fruits and nuts, medicines, building materials, fibres, waxes, resins and so on. These so called 'non timber forest products' are essential for the communities concerned and may also be a useful source of cash income if they can sell some in local markets. More recently, as we shall see later forests have been recognised as important for sequestering carbon dioxide and the question is now arising as to whether it should be treated in the same way as non timber forest products, where local communities' rights are at least partially recognised

Transaction Costs

In 1997 the Kyoto Protocol was adopted with great fanfare as a major step towards dealing with emissions of carbon dioxide, though it was not until 2005 that it came into force, having been endorsed by enough countries to become effective but it has never been endorsed by the USA. The protocol set specific emission reduction targets for those industrialised countries named in an Annex. It contained provisions for what is termed the Clean Development Mechanism (CDM) whereby the industrialised countries could partly offset their emissions, in order to achieve their targets by in effect trading emission reductions with developing countries. It also included provisions for Afforestation and Reforestation to be considered as legitimate mechanisms for sequestering carbon to offset emissions from industrial or other sources. Foresters initially saw this as a mechanism for channelling lots of money into reforestation, by persuading big emitters in industrialised countries to invest in tree planting in a developing country, that would sequester a lot of carbon dioxide, and hence reduce their net total emissions. Unfortunately it didn't turn out that way.

There are two main problems that have contributed to very disappointing results for CDM for forests and tree planting. The first is the definition of what is meant by Afforestation and Reforestation; the former applies to land that has not been forest for more than fifty years, and the second can only be used on land that was bare of trees in December 1989 (sixteen years prior to the time when the protocol came into force). The second is the need to have independent certification of the amounts of carbon sequestered. This requirement is perfectly reasonable, but it is expensive to comply with in practice. It costs little more to inspect and certify a thousand hectares as it does for ten hectares, since most of the cost is in mobilising the inspector and dealing with the paper work. The result is that CDM for tree planting is only feasible for relatively large areas in order that these costs can be spread over large quantities of carbon. It is very difficult to find large areas that comply with the definitions mentioned above, and even if there are such areas, in most developing counties there are questions over who has title to the land. It may be state land, but it is quite likely to have been occupied for many years by local people or migrants, without any legal or formal title to the land. There are then huge costs involved in sorting out who has legitimate rights to land and is willing to participate in tree planting schemes and how the benefits will be shared among all the participating stakeholders. By the time this has all been sorted out there is not enough money in the carbon that will be sequestered, to cover all the costs. As we will see later, one such proposed project in Vietnam required about 300,000 hectares to be planted for the revenue from the carbon to be enough to cover all the costs and give the farmers involved an income comparable with growing the wood for pulp or timber.

Poor Advice for Indonesia

In 1997, the 'Asian financial crisis' hit Indonesia. The Indonesian currency, the Rupiah, collapsed and the huge dollar denominated debts of both government and private sector could not be serviced. The country urgently needed short-term loans, and the International Monetary Fund (IMF) duly arrived on the scene. A couple of years before the crisis, the World Bank had been trying for some time to pressure the government into making some important reforms in the forestry sector by attaching conditions to proposed loans. However, having published a very critical report, they were told that Indonesia would not borrow any further funds for forestry. The bank staff concerned packed up and returned to Washington DC. When the financial crisis blew up, the World Bank saw its chance to use the IMF emergency funding to get the reforms that it had failed to achieve earlier. At the end of a couple of weeks of negotiations, the enfeebled President Suharto signed a Memorandum of Agreement with the IMF which included seven reforms affecting forestry among the conditions for receiving emergency loans. Although the World Bank used the opportunity to push its earlier agenda, focussing on the forestry sector was also justified by the fact that a very high proportion of the foreign debt incurred by the private sector belonged to companies in the forest product business, mainly pulp and paper.

The main thrust of the World Bank's case was that the nation's forest resources were being plundered and wasted by various cronies of the President and the government was getting little badly needed revenue that could be used to relieve poverty and promote development. This was certainly true, as many studies had shown the magnitude of the 'excess profits' that companies with forest concessions were making. Concessionaires paid a very low royalty on timber harvested based on what they used in their factories. This encouraged low levels of recovery in the forest and enormous wastage since nothing was paid for logs not utilised. In addition, they were required to pay a more substantial amount into a 'reforestation fund'. When the fund was introduced in the early 1980s, as mentioned earlier, it was intended as a deposit, which would be refunded when reforestation had been carried out successfully, but most companies found it cheaper to pay the fee and forget the reforestation. The fund grew over the years, and some of it was used by the forest department to fund a range of activities not included in the government budget, but a substantial part was used to fund activities not related to forestry at all, such as developing a medium range passenger aircraft. The World Bank wanted the reforestation fund and royalties to be abolished and replaced with a resource rent tax, which should reflect the economic value of the resource. They also proposed the introduction of 'performance bonds' by concession holders, as a means of promoting sound forestry practices. What they didn't state clearly was how the tax rate and the size of the bond should be determined and the basis upon which they should be charged in order to achieve the objectives.

In the author's book, *Making Forest Policy Work* (2002), based on detailed analysis undertaken in 1998, the implications of the World Bank's proposals are described in some detail, and it shows how the tax and the bond are interlinked and the rates must be set with great care to achieve the desired effect. There is only a certain size of cake to be shared between the concessionaire and the government, and the latter's share is divided between the resource rent tax and corporate tax on profits, while the performance bond is actually deferred profit, assuming good performance. Depositing the bond carries a cost, either in interest foregone or in the form of an insurance premium and so becomes part of the operating costs. None of these critical issues were considered by the Bank, and the government was left to work out for itself what to do. Fortunately there were a number of other donors supporting the Ministry of Forestry, including the programme that I was coordinating, who were able to help find the way through the maze.

While all this was going on, demonstrations outside the National Assembly Building, next door to the Ministry of Forestry, were increasing in frequency and strength, often making it difficult to get in to the office. Then, one day rumours began spreading that some students had been shot at Tri Sakti University, not far away in the other direction. From our windows on the sixth floor we could just see lots of people running and police firing tear gas. The following morning the papers confirmed the rumours, and the demonstrations grew larger and more vociferous. A lot of foreigners left the country, and there were stories of some being stopped on the highway to the airport by gangs and relieved of all their money and valuables. Then a few days later, one of my Indonesian colleagues came in and said there were a lot of fires. We went up onto the roof of the building, about fifty metres up, from where we could get a panoramic view of much of Jakarta. In all directions, as far as the eye could see there were columns of black smoke, which, as they rose were being slowly carried by the wind and merging into a thick blanket over most of the city.

Large parts of the city were closed off by the military, including the area around the National Assembly, which included the Forest Department, so that getting home was impossible. Fortunately, the cordon the military had established, which included the main TV station and transmitter, also included one hotel, which had few guests, and were offering special 'riot rates' on the executive floor, for the night. By nightfall, about twenty guests had checked in and we all assembled in the lounge on the executive floor, which had a huge window giving a panoramic view to the east and north, looking over the Senayan Park and sports centre. In the dark we could see the fires burning in many offices and shopping malls, and in the foreground the park was filling up with military vehicles and tanks. Awaking at about 5 am, at first light, the view from the window was of a deserted city. Most of the fires seem to have either gone out of their own accord or to have been extinguished, and the military seemed to have dispersed, so I decided it was time to try to get home. Outside, the streets were deserted, and the military cordon was no longer operating, so the short drive to the toll-way through the city was quick and uneventful, and there was no-one manning the toll gates, which were open, so the journey home took about a third of the time that it would have taken under normal conditions and for a change was gratis.

After a few days things settled down and foreigners slowly returned, and the Ministry of Forestry began to take seriously the various conditions set out in the Memorandum of Agreement. My colleagues and I began discussing the implications of the Agreement between ourselves and with officials from the Ministry and eventually produced a series of reports giving detailed recommendations as to how the conditions could be implemented. Over time most of them were gradually adopted, but our recommendations were somewhat different from those being proposed by the World Bank.

Poverty and Climate Change (2002-06)

The Poor are Discovered

The poor have always been with us, but only in the mid to late 1990s did poverty reduction become a political mantra, mainly because the Millennium Development Goals put it high on the Agenda, with targets for the reduction in levels of poverty. It is easy to set such targets but much more difficult to achieve them in practice, and we are finding that reducing poverty has many implications for the environment. The causes of poverty are manifold, especially in rural areas, often starting with a lack of education that results in the communities being largely excluded from all the information that they could use to improve their circumstances. In one village where we were enquiring about livelihoods, it turned out that most of the boys went to primary school but few of the girls. The reason given for the fact that the girls did not go to school was that in order to get to school, the children had to walk some distance which took them through a 'spirit forest', a sacred area where ancestors are buried, and they were frightened of the spirits, but the boys could run faster than the girls and so got through to the school, but the girls turned round and ran home. This was not of too much concern for the parents as the girls could help with the household chores. This particular community, like many others relied on subsistence agriculture and practised shifting cultivation. The yields of their staple crops were extremely low, so that they often had insufficient stocks to carry them through until the next harvest and so went hungry for several months each year. They also lacked a clean and reliable water supply and good sanitation.

This particular community was fortunate in many ways, because a multi-national pulp and paper company wished to establish plantations in the general area and made great efforts to involve the community. Until the wars in Laos and Vietnam in the 1960s the area had been mainly covered with a semi-evergreen forest with relatively fertile soils compared with those in the drier lowlands. During the war the Ho Chi Minh trail, which the North Vietnamese used to transport all their supplies for the fight in the south, passed through this part of Laos and the Americans bombed it mercilessly. The map below shows the distribution of the bombs, with each red dot marking the point where a bomb was dropped and it is clear from the map, that most of the bombs were targeting the roads, but plenty still landed in forests and open ground away from the roads, so that if plantations are to be established, these bombs have to be removed first. Prior to the arrival of the company many local people supplemented their meagre income by digging up the bombs and selling them for scrap, but this is a very dangerous operation, and many have been killed or maimed in the process.

When the company identified the area as being suitable for plantations they undertook to start with some modest pilot trials and to work closely with the local communities to select the areas for planting. They used satellite imagery to prepare detailed maps of the area coming under the authority of the village administration and with the help of the community members identified all the forest and other land that the community needed, including 'spirit forests' and forests for community timber production as well as land for agriculture. They then negotiated to use land not required by the community for their plantations, which would be established over a seven-year period. The company undertook to remove all the bombs and to plant the trees at one metre spacing in rows ten metres apart, and cultivate the land between the trees, which the villagers would be allowed to use for growing their crops. Each participating family was allocated a hundred metre strip between the rows of trees, and because of the lack of bombs and the cultivation, their yields were much higher than previously. As a result, all the families in the community have given up shifting cultivation and can now meet their subsistence requirement from their plots in the plantations. In addition they have cash income from the company for planting and tending the trees. Since the trees are planted over a seven year period, the families can move their plots each year so that they are always among newly planted trees which do not interfere with their crops. They can also grow some other crops among the trees during the second and third growing seasons. It is to be hoped that the company's efforts are rewarded with successful plantations, as that would bring some degree of prosperity to the local communities. This is a particularly good example of how forestry activities can improve the lives of poor people in rural areas, but alas such ventures are very few and far between.

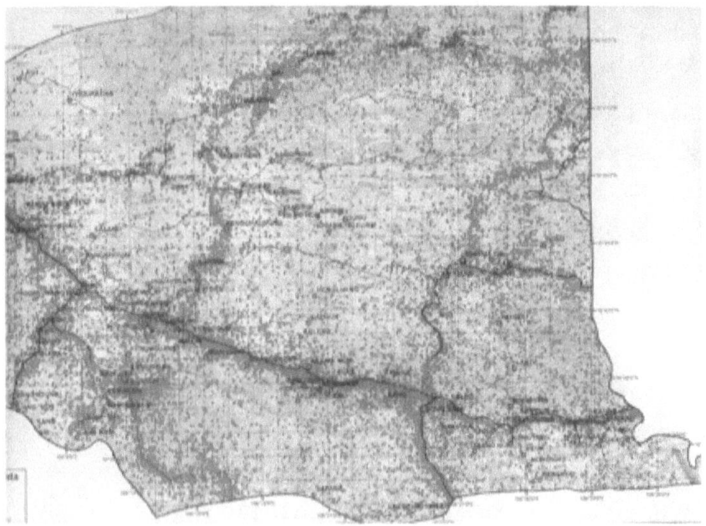

Figure 28. Map of Unexploded Ordinance sites in Xepong District, Savannakhet in southern Laos(each red dot is where a bomb was dropped according to USAF records)

Across the border in Vietnam in the region known as the Central Highlands, the ADB agreed to help design and fund a large project to support forestry and livelihoods for the local communities. It was eventually approved in 2008 and work began soon after. It was given the title Forests for Livelihood Improvement in The Central Highlands (FLITCH). The project had three investment components; the first was to support sustainable management of most of the natural forest in the region comprised of six provinces; the second was to support household and community plantations and agroforestry and some commercial plantations; the third was small-scale rural infrastructure such as water supply and local access roads. The central Highlands region lies to the west of the main Annamite mountain range and the rivers flow directly into the Mekong. Most of the region is at relatively high elevations and is home to a large number of ethnic minorities. During the war with America they tended to side with the Americans and so they have been somewhat neglected by the government in favour of in-migrants from lowland parts of the country.

During the 1980s and 1990s, as Vietnams economy started to recover from the war and grow, many poor people from the lowland areas migrated to the Central Highlands to seek land and to grow cash crops, especially coffee. This eventually led to a coffee boom and then bust as the crops matured and flooded the market with inferior *Robusta* coffee. The FLITCH project focussed mainly on the poorest Communes with a high proportion of ethnic minorities and made provision to support women especially those who were heads of households.

The project finished in 2016 and as part of the Project Completion Report, I was asked to re-analyse the economic assessment that had been made during project design.

The Project was classified as "less than successful" for a number of reasons, one of which was because the Vietnamese government ceased maintaining the monitoring for the Management Information System after the international consultants departed and so it is difficult to really know what was achieved. The available data from various sources indicates that the first component for sustainable management of the natural forest had been a bit of a failure, since the area and growing stock in the region appeared to have declined over the life of the project. This may have unknown environmental costs rather than the benefits anticipated during project design. The second component was in some ways more successful as there was a boom in both small roundwood prices and in the price of, black pepper and cassava, which were the main crops planted by the villagers. This gave the villagers a very good income, though there are signs that a bust is on the way as prices are now falling. The crops mentioned above, were meant to have been grown under agroforestry systems in alleys between rows of trees, which were intended to help reduce the risk of soil erosion, especially where cassava is planted. From the information available it seems that this was not done, and so there may be environmental costs if there is serious soil erosion when the next typhoon strikes; which almost certainly will happen one day.

Many poor communities live in very mountainous areas that are not suitable for any form of commercial plantations. In the Philippines the northern part of the island of Luzon is not only mountainous, but is also frequently subjected to torrential rain from tropical cyclones. The Philippines is struck by an average of twelve such cyclones each year, and at least one will pass over the Cordillera mountain range in Luzon. To add to the problems the region is also subject to earthquakes from time to time, which destabilises the soil and together with the heavy rain causes landslides.

From a forestry standpoint, the implication is that if the developing countries expect to achieve similar levels of consumption, as do the industrialised countries, there are big problems. The average person in an industrialised country uses the equivalent of about 0.7 cubic meters of logs annually, for construction, furnishing and particularly paper. In the developing countries the average consumption is about 0.15 cubic meters per person annually or about a fifth of that in the industrialised world. In addition the total population of the developing countries is almost five times that of the industrialised countries, so what would happen if all those people were to increase their consumption of wood? There certainly would not be enough trees in the world to provide everyone with that. At present the annual consumption of wood for industrial purposes, and forget about firewood for the moment, is split roughly equally between industrialised and developing countries, but if all the people in the developing countries increased their consumption of wood to the same level as in industrialised countries the total requirement for logs would increase more than threefold.

Coordinating Development Assistance

During the course of preparing the new project in Laos to promote investment in tree plantations for poor farmers mentioned above, discussions had been held with the Japanese International Cooperation

Agency to seek a contribution to the project in the form of technical assistance. The head of the JICA office in Laos had given an indication that, in principle, Japan was interested to provide a couple of experts for the project. During the appraisal mission for the project the team from ADB visited the JICA office to discuss the details of the project and the form of technical assistance that would be most appropriate. The ADB mission consisted of three persons, and the Head of the JICA office had two of his assistants present, so that there were six people seated round a table. The leader of the ADB mission sat next to the JICA Head. After the introductions and an explanation of the purpose of the visit by the ADB team leader, the Head of the JICA office responded.

It immediately became clear from body language that he had a difficult message to relay. He spoke at considerable length in a voice so quiet, that even the ADB team leader sitting next to him, could hardly hear what was being said. As far as could be understood, all the JICA funds for the immediate future had already been committed, and the priority for future funds was to help the government with the preparation of a forestry sector strategy. Although it was explained that part of the ADB project would be the preparation of a long-term strategy for the development of plantations in the country it became clear that JICA had already committed all its funds to other activities and was no longer in a position to support the ADB project. This illustrates the difficulties in coordinating development support between several different donors, who each have their own procedures and schedules and find it extremely difficult to synchronise.

Climate Change and REDD+

It took some time for climate change to rise up the political Agenda after the UN Conference in 1992 that led to the establishment of the UN Framework Convention on Climate Change (UNFCCC). One of the outcomes was the spawning of a huge array of new acronyms; COP, GHG, LULUCF, REDD, CC, CDM, IPCC and many more, but it eventually led to the Kyoto accord reached 1997, that eventually came into force in 2005, and is the first international agreement that attempted to promote some specific actions to reduce the global emissions of carbon dioxide and other Green House Gases (GHGs). This growing awareness of the possible significance of the role of various gases, including carbon dioxide in the heat exchange mechanism in the atmosphere, was something that took me back about forty years to the time when research into primary production was an important topic. In the mid 1960s it was already apparent that carbon dioxide concentrations in the atmosphere were increasing, but at that time it was generally thought that a small rise in temperature and increase in carbon dioxide concentration would increase the growth rate of many crops and plants, and especially trees, would mop up most of the additional carbon dioxide so that nature would solve the problem.

Nowadays, Governments from developing countries, when asked to reduce their emissions of greenhouse gasses (especially carbon dioxide) understandably take the view that why should they cut back on energy consumption to reduce emissions, when most of the increase in carbon dioxide in the atmosphere has come from the industrialised countries. Most feel that they need to industrialise too in order to reduce poverty. In fact there is another side to this argument, which we will come to later.

As mentioned earlier, the Kyoto Protocol included the Clean Development Mechanism (CDM), which was intended to facilitate measures to reduce emissions of Greenhouse gases and also to promote measures to sequester carbon dioxide. Forests sequester and store substantial amounts of CO_2 and so some governments and forest departments saw this as a way of attracting funds for afforestation and reforestation from major emitters of carbon dioxide to offset against emissions that went over the limit set for them by their government to meet targets agreed to at Kyoto. The early optimism was soon dashed as proposals ran up against a number of practical problems, including the high transaction costs discussed earlier. If a company that emits large quantities of carbon dioxide is going to pay for a certain amount of the gas to be sequestered by trees it needs some guarantee that the agreed tonnage of carbon dioxide has actually been locked up in the form of cellulose. It also needs to be able to demonstrate to the authorities that it's actual emissions, minus the amount sequestered, is within the limit that it has been allowed.

Further problems arose, because the requirements also dictated that the trees planted to sequester the carbon dioxide would not otherwise have been planted without the payments for offsetting emissions, a feature know as 'additionality'. If the trees would have been planted anyway, there would not have been any net additional sequestration. There was also a requirement that the land where the trees were to be planted had been devoid of trees since 1989, to avoid the possibility that people cleared forest in order to get the funds for re-planting. In developed countries it was not too difficult to meet these requirements, because historic records of land-use are generally available, and plans for tree planting are normally incorporated into some form of management plan, making it fairly easy to show that proposed planting is additional to what would have taken place. There is also secure right of tenure, either through ownership titles or lease agreements for land in developed countries, so that it is more or less guaranteed that trees planted will remain undisturbed and sequester carbon dioxide for many years. Techniques and expertise for measuring the carbon content of the planted trees from time to time are readily available.

In most developing countries, none of these conditions hold true; it is usually difficult to find satellite images of an area where trees could be planted to prove that it was deforested before 1989, and if by chance some cloud free images are available, the ownership over the land is usually uncertain. It may nominally be state land, but is often occupied by people who have just moved in, or by people descended from communities that have lived in the area for generations, but whose rights are not recognised or have just been usurped by the state. With such confusion over property rights it is perhaps not surprising that most people living in remote rural areas have little knowledge of the niceties of legality and so tend to use land or whatever they find growing on it. Throughout most of the humid tropics forests dominated the landscape until people moved in. While it may be possible to find land that was cleared of forest before 1989, it is rare to find such land that has been abandoned and is therefore available for planting with trees. Such abandoned land, where it can be found, is usually seriously degraded and has become grassland, that is both costly to re-establish with trees and is infertile so that tree growth is slow. All these difficulties mean that finding projects that qualify for funding under CDM is difficult and time consuming, which tends to make the 'transaction costs' high, and this effectively excludes small projects that will only sequester a small amount of carbon dioxide. In Vietnam, in 2004 a feasibility study for a possible CDM project to plant trees in the central highlands as part of a larger project for improving livelihoods through tree planting and other activities, found that at least 300,000 ha of special carbon dioxide sequestering plantations would be needed in order to spread the start-up and monitoring costs over a big enough area to keep the cost per tonne of carbon dioxide down to a level where the project would be profitable with the prevailing carbon price. Needless to say the project did not go ahead.

As a result there have been very few successful CDM forestry projects in developing countries, and there is now, understandably a fair degree of scepticism among forest departments in developing countries that they can benefit from climate change related funding. In the meantime data on the various sources of emissions of carbon dioxide was being extended and developed, and it became apparent that the clearance of forests for conversion to other land-uses and the steady degradation of the remaining forests due to over-logging was contributing very substantially to the overall global emissions of carbon dioxide

The Fourth Assessment Report of the Intergovernmental Panel on Climate Change in 2007 estimated that emissions of all green house gases from forests and forest land-use changes, mainly in tropical countries, accounted for almost twenty per cent of global GHG emissions. The Stern Review[3]in 2006 had concluded that one of the least cost ways of reducing emissions of GHGs would be to pay tropical developing countries to protect and retain their forests rather than clearing them and converting to other land-uses. It did not recommend any particular figure for reducing deforestation but pointed out that halting deforestation would be just one measure necessary to stabilise atmospheric carbon dioxide concentrations at 450 parts per million. The review does, however, support the principle that payments for protecting forests should be calculated according to the 'opportunity cost' of the use of land that would otherwise be available for agriculture. This means that the 'lost profits' from not converting forest to oil palm plantations or agriculture, would have to be paid. However, it did not indicate how to deal with a situation where the 'opportunity cost' of not converting the forest exceeds the value of the emissions avoided at the prevailing market price for carbon.

---

[3] "Stern Review on the economics of climate change", 2006, final report, http://www.hmtreasury.gov.uk/independent_reviews/stern_review_ economics_climate_change/stern_review_report.cfm

In 2005 the eleventh Conference of Parties (COP11) set up a working group on reducing tropical deforestation, which reported back at the end of the 2007 to COP 13 in Bali. It took account of the growing consensus that deforestation must be tackled in order to reduce total greenhouse gas emissions, and recommended that the cost for doing so must be shared between nations and not be borne by the, largely poor, rainforest nations. Those UNFCCC discussions were initiated by proposals submitted by Costa Rica and Papua New Guinea and then endorsed by a large number of countries and institutions, called the Coalition for Rainforest Nations. Those proposals were based on Costa Rica's 'Payments for Environmental Services' scheme, which involves payments for 'those services provided by forest and forest plantations to protect and improve the environment'. They referred to the concept as 'Reduced Emissions from Deforestation in Developing Countries' (REDD) or 'Avoided Deforestation'.

The stated aim of the Coalition for Rainforest Nations was "to incorporate certified emissions offsets related to deforestation (in addition to afforestation and reforestation) within global carbon emissions markets by revising the Marrakech Accords, amending the Kyoto Protocol, or developing a linked 'optional protocol' under the UNFCCC". At present, carbon trading provides funding for 'afforestation and reforestation' but under the Kyoto protocol, no carbon finance is available for protecting existing forests. The Coalition for Rainforest Nations aims to raise large sums of funds for conserving tropical forests, much of it via carbon trading mechanisms. Their proposal submitted to a REDD workshop in March 2007 aimed at a combination between emissions trading and a separate fund. New funding would be for protecting existing forests only, though Clean Development Mechanism funding for afforestation and reforestation would also be expanded. The trading mechanism would be introduced after 2012 and would be in addition to emission reduction targets by Annex 1 nations. It was expected that the details of such a scheme would be negotiated and agreed at the Copenhagen Conference (COP15) in December 2009. There were a number of other proposals put forward including 'Compensated Reduction' as proposed by India; Brazil's proposal for a non-market fund for reducing deforestation emissions, without any obligations on developing countries to reduce emissions, and Tuvalu's Forest Retention Incentives Scheme, which would be an international fund for community-based forest management schemes.

None of these proposals really addresses the root causes of deforestation and degradation nor the problem that the global area of forest that can capture and store carbon has declined to the point where it can no longer keep up with the increase in emissions.[4] Forest destruction is driven by a combination of poverty, whereby farmers with no valid land title clear forest to increase food production and incomes, and by economics, whereby much land has a far higher commercial value for growing almost any crop, than it has as forest when just timber values are taken into account. The economic value of forests, however, in the form of soil, water and biodiversity conservation and other intangible benefits may be quite high, but is difficult to determine, because these services that forests provide are rarely marketed and so they have no direct monetary value; they have to be estimated by various indirect means, such as by asking people how much they are 'willing to pay' for the service. The value of forest for water conservation could be estimated by looking at the likely cost of flood damage if the forest in a river basin is cleared and converted, but it is not a straightforward relationship. This gives forests in river basins that are well developed with, for example, hydro-power dams, irrigation schemes or other infrastructure or habitation, much higher values than forests in underdeveloped river basins. During an investigation in the Philippines for ways in which REDD funds might be applied to reduce deforestation in the Cordillera mountains in northern Luzon, it was found that many farmers were clearing forest on very steep hillsides to grow vegetables. For many of them this was a very profitable business, and the value of the land for the vegetables far exceeded any carbon value of the forest that was being cleared. Although forests provide many environmental services that are beneficial to the community at large neither a poor farmer or a rich farmer or corporation growing oil palm, bananas, pineapples or other such crops pays the economic cost of losing these services nor are they rewarded for providing such services.

The variety of proposals for dealing with carbon dioxide emissions from forests illustrates the complexity of the situation and these were

---

[4] 'Reduced Emissions From Deforestation': Can Carbon Trading Save Our Ecosystems? By Almuth Ernsting and Deepak Rughani July 2007

reflected in the discussions in Copenhagen, Cancun, Durban and Warsaw and Paris, with so far, no clear agreement reached as to what will be funded and how. Under the original REDD concept, payment would only be made for reductions in emissions which would mean little or no reward for countries that have been successfully protecting their forests, since they would have a hard time proving that they had reduced emissions. Similarly, a country that had lost much of its forest in the past is likely to have reached a point where the rate of forest loss has slowed and emissions are declining anyway without any interventions so that it would be difficult to prove that additional emission reductions had been achieved. It also raises the question of whether the emission reductions should be measured, and paid for, in relation to a specific forest area, or on a national or regional basis. The former position raises the difficult problem of 'leakage', which is the displacement of emissions from one place to another, while the latter makes it difficult, if not impossible to assign the payments to those that have successfully achieved emission reductions, since their efforts could be nullified by contrary action elsewhere in the country.

However, since the international protocol under which REDD payments might/will be made has not yet been finalised, it is important to continue examining and testing all possible means for reducing emissions in order to determine the real and actual cost. Such studies and pilot activities also need to see how a range of other Payments for Environmental Services can be combined so that investments in protecting and regenerating forests become not just economically viable, but also financially attractive to the private sector and the farming community. Only by trying to achieve emission reductions can the practical difficulties be highlighted and the costs and benefits be better understood. This is a new challenge for those foresters who work in the international arena.

REDD+ in the Mekong Region

The World Bank established the Forest Carbon Partnership Facility as a global partnership of governments, businesses, civil society, and Indigenous Peoples focused on reducing emissions from deforestation and forest degradation, forest carbon stock conservation, the sustainable management of forests, and the enhancement of forest carbon stocks in developing countries. At the present time, thirty six tropical developing countries are members, including Thailand, Cambodia, Lao PDR and Vietnam from the Mekong sub-region. Member countries can receive a grant of funds from the facility to prepare a Readiness Preparation Proposal (R-PP) which describes the country situation on forest resources, past and current deforestation and forest degradation, estimates of current levels of emissions of carbon dioxide resulting from land-use change from forests to other uses, and an indicative strategy for reducing emissions in the future. It also describes the institutional situation in terms of organisations, laws and regulations and human resource capacity as well as all the various stakeholders, including government departments, civil society organisations and private sector companies. The overall purpose of the document is to set out how the country will prepare itself and the funds required for implementing a REDD payments system in four or five years time, when it is expected that there will be an internationally agreed protocol along the lines of Kyoto.

An R-PP has been developed by all four countries during the period 2010-2013, which gives them access to a grant of around US$ 3.6 million as a contribution to the REDD preparation process. The country R-PPs indicated that substantially greater funds will be needed to undertake complete Readiness preparation to include piloting a range of activities to act as a basis for the REDD Strategy. One of the most contentious issues is the benefit sharing arrangements, particularly at present when carbon values and hence potential revenues are very low. Most rural communities living in and around forests will be required to contribute substantially in many ways to forest protection and restoration, but they are very concerned that much of the funds will stick to fingers along the way, if governments are involved in handling the funds. Likewise governments see REDD payments as a way of supplementing budgets for forest management and expect a substantial proportion of the revenues to flow into government coffers. It is far from clear as to how much emissions can be reduced by various measures, such as patrolling forests and clear demarcation of forest boundaries, nor how much such interventions will cost in relation to the level of emissions that can be achieved.

One thing that the preparation of the R-PPs has highlighted is the magnitude of the emissions at present from the forestry sector in each country. Both Thailand and Lao PDR had submitted National Communications prepared by the national agency charged with responsibility for climate change issues to the UNFCCC reporting on their emissions a decade or so ago, and both reported very low emissions or net sequestration from their forests. However, the more detailed and thorough analysis conducted for the R-PP by the respective forestry agencies with the best data available indicated that emissions are probably very substantial, and that degradation of the forest through illegal logging and shifting cultivation, that reduces primary forest to a much lower density secondary forest, contributes around a half of the total.

The R-PPs for each country, having identified the drivers of deforestation and forest degradation also include a preliminary strategy for tackling them with a series of measures that can be piloted to gain experience and evaluate their effectiveness. One of the most difficult drivers to deal with is illegal logging, because there are so many vested interests involved and it requires policing very large areas. In many of the countries the military is involved in illegal logging, especially in remote border areas where there is restricted access for civilians The main approach adopted by most countries is to engage local communities in patrolling the forests around their villages, and reporting any activity that they encounter. Sometimes the villagers are actually employed by those who want the logs, who are frequently foreigners in Laos and Cambodia, and often the villagers do not realize that they are doing anything illegal. When a village patrol spots something that they believe to be an illegal activity they are expected to report it to the nearest government forestry office or the police, but few patrols are equipped with radios, which don't work well in thick forest, and few are accompanied by a representative of the police or the forestry department and so it may be a day or more before the report reaches anyone who can take action.

During the preparation of the R-PP in Laos, I suggested that a good approach would be to use drones, which are now becoming readily available and reasonably priced. I contacted some of the companies that are making commercially available machines, and obtained both cost and operational information, which indicated that it would be both practical and much more cost effective than ground patrols. I also learnt that Malaysia is currently using drones to monitor inshore fishing. I thought the idea would be received with enthusiasm by the forestry department, but they said that they would have to get permission from the government and the military, which they thought was most unlikely to be granted. All references to drones in the final document were removed, but I did manage to use the term 'modern technology' in order to leave the door open for the use of drones in the future. Some funds were allocated in the budget. At the time of writing the topic is still not open for discussion, though a number of internationally funded projects are hoping to push the idea.

At the root of the problem in all four countries is the issue of land tenure. Forests have been declared as state land, and forest management units such as national parks, reserved forests, protection forest and production forest areas have been marked on maps and declared in various legal instruments, but the boundaries have not been negotiated on the ground with the local communities nor have they been surveyed and marked on the ground. Sometimes existing communities have found themselves inside a designated forest and are then accused of illegal occupation, and on other occasions communities have expanded the area they cultivate into designated forests without realizing it, since there has been little or no consultation with them prior to designation. This is a major issue that will need time and considerable funding to resolve before REDD can become a reality.

Carbon Neutral Transport Corridors

Five countries (Myanmar, Laos, Thailand, Cambodia and Vietnam) together with Yunnan, the south-western province of China, have been designated as the Greater Mekong Subregion (GMS) by the Asian Development Bank for the purposes of coordinating development. Part of the development focuses on transport infrastructure and several 'corridors' that connect major urban centres across the Subregion have been the focus of investment in bridges and road upgrading. One of these, termed the East-West Economic Corridor (EWEC) runs from the seaport of Danang in Vietnam on the South China Sea across Laos, Thailand and Myanmar to the seaport of Moulmein in Myanmar on the Andaman Sea. Another, termed the North-South Economic Corridor (NSEC)runs from Kunming, the capital of Yunnan Province, south through Laos and Thailand to Bangkok.

Bridges across the Mekong connecting Laos with Thailand have been built for the two corridors connecting Savannakhet in Laos and Mukdahan in Thailand on the EWEC and connecting Houay Xai in Laos and Chiang Kong in Thailand on the NSEC. Several sections of the roads have been upgraded to dual carriageway, especially in Thailand and more is underway. The result is that traffic has increased substantially, which means that emissions of carbon dioxide are increasing, especially as much of the heavy traffic is made up of old and very inefficient trucks. In the light of efforts to reduce emissions of carbon dioxide, an ADB staff member came up with the idea, of trying to create carbon neutral corridors, whereby emissions of carbon dioxide from vehicles might be offset by carbon dioxide sequestration by trees planted along the corridor. Such tree planting would have additional benefits such as protecting the roads from erosion in hilly areas and providing shade for travellers, especially in places where service facilities are, or could be, provided.

In 2009, I was asked to carryout an initial study to investigate whether either or both of the corridors mentioned above could become carbon neutral as far as freight traffic emissions are concerned. The initial work involved collecting data from various sources on traffic flow along sections of the two corridors, broken down by type of vehicle. Sufficient data was found on an Asian Highway Database to enable an estimate of the current emissions, and this suggested that emissions on the EWEC were about one million tons of carbon dioxide annually and on the NSEC about two million tons carbon dioxide annually. The latter was higher because it included traffic flowing into and out of Bangkok, whereas the EWEC has only small and medium sized urban areas along its length. These figures were shown to be fairly accurate by a later more detailed follow-up study.

The next part of the work required the corridors to be surveyed on the ground by travelling the length of each of them to observe the traffic and identify possible areas where tree planting could be undertaken. To help me, I engaged the services of two young Thai professionals, one an expert in computer based Geographical Information Systems (GIS) and the other a forester. I arranged in advance for vehicles to be hired or made available in each country and to meet us at the relevant border crossing according to our planned travel schedule.

For the EWEC we flew to Danang from Bangkok and met up with the local Vietnamese forestry officials and briefed them on the purpose of our visit. They came with us as we travelled up to the Lao border, stopping from time to time to take photos and record information on the local landscape. We stayed overnight near the border, and the following morning we said goodbye to our hire vehicle at the border and walked through the Vietnamese check point, across a couple of hundred metres of no man's land to the Lao checkpoint and then into Laos, where another hire car was miraculously waiting for us. By nightfall we had reached the town of Savannakhet on the banks of the Mekong and our hire car dropped us at the international bridge. Pedestrians are not allowed to walk across the bridge so we had to wait for a shuttle bus that conveyed us across to Thailand. The Thai end of the bridge is some way out of the town of Mukdahan and so we had to take a three wheeled 'tuc tuc' into town and find a hotel for the night. In Thailand taxis are relatively cheap and are regularly used to travel quite long distances, so the driver of a local taxi was not surprised when we said that we needed him for the next four days to travel across Thailand to the border with Myanmar and then take us to Bangkok, a total distance of almost 2,000 kilometres. The first part of the journey was across the Khorat Plateau, which is more of a shallow basin than a plateau, being almost surrounded by mountains that cut it off from the rest of Thailand. It is relatively dry compared with other parts of Thailand and has extensive fields of cassava. Large numbers of Dinosaur bones from many species have been found scattered all over the plateau.

In the middle of the Plateau is the town of Khon Kaen, where there is a pulp and paper mill that uses eucalyptus wood mainly from local farmers' small woodlots. The farmers were initially encouraged to plant the trees by a government scheme that gave them grants to cover much of the initial cost, but now many of the farmers have found that growing the eucalyptus trees is not profitable and so have stopped planting new areas and are clearing their old crops to grow cassava, which is more profitable. This has left the company short of raw material, which it is now trying to purchase in Laos. Not much hope to grow trees in this region for carbon sequestration.

Further west we cross the mountain range that forms the boundary between the Khorat Plateau, whose rivers flow east into the Mekong and the basin of the Chao Phraya river and its many tributaries, which flow south into the Gulf of Thailand. These mountains are generally well forested and include a couple of national parks, but after the wars in Laos many refugees from the Hmong ethnic minority, who backed the wrong side in the war fled to Thailand and settled in the area. These people like to live in upland areas and practice shifting cultivation and like to clear forest to grow a range of crops including opium poppies, so there are very substantial areas that have been denuded of trees and are now mainly grassland; maybe good candidates for growing carbon sequestering trees.

There are three rivers that rise in the mountains to the north of Thailand that flow south so the EWEC goes up and down quite a lot as it crosses from one valley to the next, but the upper parts of the mountains are still largely covered in forest, but moat seems to have been logged over in the fairly recent past and do not have many large trees anymore. There is one final range of mountains to cross before reaching the border with Myanmar, which is formed by the Salween River. At the time of the study there was no bridge over the river, and Myanmar was not included in the study for political reasons, but that has now changed. This final range of mountains is also forested and has a National Park, but like the earlier mountain areas there are large patches that have been deforested and are now largely grassland. The government has some small plantations of Khasya pine, which is indigenous in the higher areas locally. So again some scope for tree planting for carbon sequestration.

The NSEC is a different *kettle of fish*. The trip started in a similar fashion, flying to Kunming and meeting local Chinese officials for briefing and discussions. They laid on a vehicle and interpreter to accompany us to the Laos border. In the evening at dinner, I discovered that there was quite respectable locally made grape wine available. Evidently French Jesuit missionaries in the nineteenth century had brought in grape stock and set up wine making, which has been continued to this day. In the morning before departure, I bought a few bottles to provide my daily tonic during the journey

The first five hundred and forty kilometres south from Kunming to Jinghong is on a fairly new motorway running south and south-east over a highland plateau at around 1650 metres above sea level, crossing several river valleys including the Yuan, (which becomes the Red river in Vietnam) and the Pa Dien, on spectacular bridges and passing through several tunnels before descending to the Mekong (known as Lancang in China) valley. The area is densely populated, with forest confined mainly to the hills, while corn, tea and vegetables are widely planted. There are many industries and for the final 60 kilometres into Jinghong the road passes down a steep-sided valley covered in rubber plantations. These are of increasing concern to the Chinese government, because of the amount of erosion that is taking place. An enjoyable evening was spent in Jinghong, including entertainment by a troupe of dancers from a local ethnic group. The following day we travelled to the Lao border, visiting the lovely Xishuangbanna botanic garden *sen route* and traversing some fairly rugged and still largely forested terrain. The motorway was still under construction at that time and the road was barely passable in some places.

On arriving at the border, the road was lined with new buildings with offices and shops on both sides for about a kilometre, all landscaped with trees and flower beds, and we drove up to a large building with an arch in it, which straddled the road. It turned out that we had driven straight past the checkpoint and had officially left China without clearing immigration and customs. The driver made a U turn and we went back to look for the checkpoint. As foreigners we were allowed to complete the formalities after a long delay while the driver was interrogated and we left him negotiating while we walked up to the archway, through it, across a kilometre or so of *no man's land* and into the Lao check-point in a scruffy little wooden hut. The Lao formalities were relatively easy and our hire car was waiting for us more or less outside. The area around the checkpoint had been cleared for a distance of about two hundred metres in all directions and had a scattering of huts and small buildings and plenty of hawkers selling many things. At the edge of the clearance to the west there was a very large and grand new building in the final stages of construction, and we were told that it was a Casino for Chinese tourists as gambling is forbidden in China.

The first fifty or so kilometres through Laos were mainly through young rubber plantations, and we learnt that Chinese entrepreneurs come to Laos and take local government officials for tourist and shopping trips to China in return for help with persuading local farmers to plant rubber trees. The farmers who agree are given five hundred and fifty rubber tree saplings on account, each valued at US$1, which is to be paid back with interest when the trees are ready for tapping after about seven years. At the end of the first year the trees are checked and if the total number of trees living is less than five hundred, the farmer has to purchase the necessary number of trees to make up the five hundred at a cost of US$10 each, again on account. We were told that some farmers were unable to make enough profit to repay the money advanced to them, and so their land was repossessed, by the government and handed over to the Chinese entrepreneurs. Not the best way to alleviate poverty!

The rest of the journey across northern Laos took us through a large National Conservation area, which is subject to some unauthorised clearance, but has a number of very successful 'ecotourist' facilities, including a large 'Tree-house'. The area is very hilly and is now generally denuded of forests and much has become grassland. Another good candidate for carbon sequestration! Before reaching the Mekong at Houay Xai we passed a large open cast lignite mine which has resulted in a vast area of forest being cleared and a huge hole in the ground now filled with water. The lignite is exported to Thailand to fuel an enormous power station.

The Mekong is the international boundary between Laos and Thailand and we cross it in a small long boat with an inboard motor, through the Thai immigration on landing and by 'tuc tuc' to a local hotel on the bank of the Mekong for the night. In the morning our hire car turned up and we set off. The rest of the journey to Bangkok took three more days and mostly followed the river valleys, so less up and down but quite a lot of somewhat degraded forest.

The final fifty kilometres into Bangkok is one big traffic jam and no hope of planting any trees, but overall there seems to be sufficient land, not obviously being used for any productive purpose, which could be afforested to sequester enough carbon dioxide to more than offset that generated by the traffic in the corridors, so our report results in a follow-up more detailed examination of the feasibility of the concept.

Wood and Silicon Chips

Travelling in the island of Mindanao, in the southern Philippines, it was a bit of a surprise to learn that several trucks loaded with charcoal were leaving a village every day for destination unknown somewhere in the north of the island. Large quantities of charcoal were being made in the village using primitive earth kilns to convert a mixture of wood cut illegally from the remnants of natural forest in the vicinity and wood from trees planted around the village. Since the visit was to examine the causes of emissions of carbon dioxide in the area as part of an assessment of the total emissions from forestry activities it was clearly important to find out more about where the charcoal was going and how much was being used every year.

Enquiries eventually revealed that the charcoal was going to a factory near Cagayan de Oro that was using it to make silicon. The most commonly used source of silicon is quartz, which is silicon oxide, and to convert it to silicon it must be reduced by heating it with carbon, that removes the oxygen from the quartz to make carbon dioxide and leaves pure silicon, which can then be used in computers, solar panels and other electronic devices. This particular factory was using about 30,000 tonnes of charcoal annually, which was delivered by about ten trucks each carrying about ten tonnes every day, plus small amounts delivered by local farmers in pick-ups.

The factory had four large retorts powered by electricity; the quartz and charcoal were fed in at the top and mixed as they descended slowly through the retort with molten silicon emerging at the bottom and carbon dioxide being vented from the top. An interesting question was to which sector the emissions should be assigned and should the emissions from the charcoal making process be considered as a contribution by forestry or by industry. It has subsequently appeared that there are similar plants in Thailand and Laos with the latter mainly supplying silicon to China. In nearby Yunnan Province of China, there are reported to be three such plants, one of which uses charcoal made from bamboo

Figure 29: Raw materials used in manufacture of silicon; charcoal (left) and quartz (centre) and the result pure silicon (right)

Forests and Coal

The Indonesian part of the island of Borneo, known as Kalimantan is rich in forests above ground and coal below ground. The coal below ground is not at great depth and so is normally mined by opencast methods that involves clearing significant areas of forest. Although the regulations for mining require the mining companies to restore the forest, inevitably, if the forest is restored, it is a monoculture and in no way replaces the biodiversity of the forest that is removed. It is ironic that the world's politicians and environmentalists are putting so much emphasis on reducing or preventing deforestation as an important measure for reducing emissions of carbon dioxide and doing nothing about reducing emissions resulting from coal mining.

In one coalmine in South Kalimantan a company has a concession to mine coal over an area of about 2,600 ha and currently is opening up about forty five hectares every year to produce about 3-3.5 million tonnes of coal. This represents almost 80,000 tonnes of coal per hectare or eight tonnes per square metre. This is fairly consistent with the average thickness of the coal seams being mined of about five metres. The carbon content of coal varies considerably and in the South Kalimantan mine was reported to be about forty five per cent so that one hectare of coal would contain about 36,000 tonnes of carbon, which, when it is burnt will produce more than 100,000 tonnes of carbon dioxide with a current value of around US$ 500,000. This can be compared with the roughly two hundred tonnes of carbon that was probably in the forest removed, giving carbon dioxide emissions from a hectare of the forest that is cleared of about seven hundred tonnes with a value of around US$3,500. Is it any wonder that coal companies are not interested in reducing the emissions by avoiding clearing forest !

Figure 30: Open-cast coal mine under former forest in South Kalimantan, Indonesia

A hectare of planted forest using indigenous forest species will sequester about five tonnes of carbon every year, using around eighteen tonnes of carbon dioxide, so that it will take about forty years to sequester seven hundred tonnes carbon dioxideand accumulate the two hundred tonnes of carbon in the original forest at a cost of around US$1,000. At this rate about 5,500 ha of plantations would be enough to mop up all the carbon dioxide produced by burning the coal removed from under one ha of forest every year. These numbers stack up, since the value of the annual emissions from the coal would be enough to fund the establishment of about five hundred hectares of plantations, and only eleven years of planting would be needed to offset all the emissions. In the case of the South Kalimantan mine, all future emissions from burning the coal extracted each year would be offset by planting a total of about 250,000 ha. Although this area is about a hundred times the area that the mine will exploit, it is enough to restore all the degraded forest in the surrounding area within the forest management unit (KPHP) declared by the government.

Other mines will have different local conditions, but the principle seems to be valid, that coalmines in forested areas should offset the emissions from burning the coal by restoring an appropriate area of degraded forest. Responsibility would be on the user of the coal rather than on the mining company, unless the coal price is adjusted to include the cost of the subsequent emissions. It would be nice if this concept could be put into practice, but it seems unlikely that power utilities and power consumers would be prepared to offset the environmental cost of the emissions resulting from their use of the fossil fuel.

Full-Circle: Carbon Dioxide and Airships Again

Following the COP 13 in Bali in December 2007, mentioned above, the German government signed an agreement with the Indonesian government to fund a wide range of pilot REDD activities in order to gain experience for the future. They were also intended to provide information that would better inform the up-coming negotiations, that were expected to, but in the end did not, culminate in Copenhagen in December 2009. One result of this was that I found myself in Indonesia again with a team of consultants charged with coming up with a detailed proposal on how to save the forests. It was eventually agreed that the investigation should focus on East and West Kalimantan especially the Districts in the north where the two provinces meet up and also share a border with the Malaysian States of Sabah and Sarawak. The governments of Indonesia, Malaysia and Brunei have signed an accord to establish a network of protected areas on both sides of the borders to be called the Heart of

Borneo. On the East Kalimantan side in the District of Malinau there is the Kayan-Mentarang National Park, which extends to about 400,000 hectares and is very remote. It is either a three-day trek by canoe up one of the rivers, or a fairly expensive flight in a small mission aircraft, to get into the heart of the Park. The terrain is rugged and being mainly covered with fairly dense forest can only be penetrated along the rivers. There is a plan to build a trans-Kalimantan road connecting Samarinda on the east coat with Pontianak on the west and this has prompted plans for feeder roads, one of which would pass through part of the Kayan-Mentarang National Park. Many people are opposed to any road building in the area as once access was improved, prevention of further large-scale destruction of the forest would be almost impossible.

There are a few small Dayak communities that live within the National Park, and the Park management are hoping to support their livelihoods by promoting a range of economic activities. Several community members have been trained as guards and are paid to carry out foot patrols through out the Park. On satellite imagery it can be seen that the forest on the Malaysian side of the border has been logged and there is a dense network of roads, which come right up to and even over the border into Indonesian Kalimantan. Some Malaysian loggers were caught on the Indonesian side of the border, and had all their heavy equipment confiscated, but it is clear that the forest is under considerable threat. There are also companies, several of them Malaysian, seeking to clear forest and establish oil palm plantations on the slightly flatter land in the east of Malinau District, and there is one open-cast coal mine that is clearing forest. The latter are doing some restoration by planting fast growing trees, but it will be a long time before the carbon stocks in the natural forest that is being cleared, are replaced if ever.

Each of these pressures on the forest requires a different solution and perhaps the most challenging is the development of the National Park without causing the destruction of the forest. If it is to function as a National Park that can be enjoyed by Indonesians and foreign tourists, improving access is essential, as few tourists, especially middle-aged ones, are likely to welcome the idea of a long trek, even if overnight accommodation is created at staging posts along the route. With the Park also threatened by illegal loggers, mainly from the Malaysian side at the moment, it is essential to mount regular patrols to check what is going on and apprehend culprits. If land access is improved on the Indonesian side it will almost certainly result in illegal logging and encroachment with clearing forest by landless migrants from other parts of Indonesia.

A solution to this challenge takes us back to Chapter VI when we were looking at transporting charcoal in Nigeria, and one answer was to use airships. It seemed that this could be a perfect solution for Kayan-Mentarang National Park since tourists could be transported relatively quickly to sites within the Park, produce from the local communities within the Park could be transported quickly to markets, and the Park Managers could mount regular air patrols of the Park, all without the need to build any roads. Some enquiries revealed that although the company Airship Industries, that was building and developing airships in the UK in the late 1970s had been sold out to a Canadian company there is a new company currently developing commercial airships in the UK, and contacts with them were fairly positive, in that they were interested in the concept and thought it possible that although their current plans are for a large airship, they could be interested in a smaller version that could meet the needs of, not just Kayan-Mentarang National Park, but possibly many other Parks throughout the world with similar problems. Strangely, just as I am writing this, a picture appeared in my local newspaper showing their aircraft the Airlander 10.

Experience in designing REDD investments in Indonesia during 2008 has shown that the current market value of carbon dioxide emissions avoided or sequestered at around US$ 5 per tonne is insufficient to compensate those responsible for deforestation and degradation for the opportunity cost of continuing 'Business as usual'. Studies in Philippines by ICRAF have come to the same conclusion as they have shown that the abatement cost of carbon emissions through forestry is much higher than this current projected price for carbon dioxide.

After the Wars in Cambodia

Anyone wanting a detailed account of what went on in Cambodia leading up to, during and in the immediate aftermath of the Khmer

Rouge period, should read Elizabeth Becker's book *When the war was over*, from which it is clear that the Cambodian people endured horrendous conditions for the four years of the Khmer Rouge's period in power. While there are no longer the paranoid killings and general disruption to life due to the constant forced movement of people from one part of the country to another, there are other issues that make life difficult for the ordinary person.

Compared with the other countries that lie within the Mekong river watershed, Cambodia still has substantial forests. In fact during the wars these forests helped greatly to sustain the people by providing basic necessities including food, fuel, materials for constructing shelter and many things needed for daily life such as utensils and tools and other farming equipment as well as some income from selling various products. As in the other countries in the region the population is made up of many ethnic groups. In each country the lowland groups have become dominant and those groups that traditionally lived in the uplands have generally become marginalised and are referred to as ethnic minorities by the dominant lowlanders. Most of these highland groups practiced shifting cultivation and have traditions for managing the forest that bans indiscriminate cutting of trees to ensure the sustainability of the forest and also includes the recognition of certain areas of forest as 'spirit forests' and others as being of special importance for particular non timber products, such as medicines, fibres and so on. These areas are sacred and tradition has it that cutting down any tree in a spirit forest will have dire consequences.

Throughout the lowlands of southeast Asia from eastern India to Indonesia the dominant family of trees is the Dipterocarp. Species from this large family of trees are found in many forest types. There are hundreds of species in the family and most are prized for their timber, which ranges from almost white, though yellow, pink and various shades of red to dark red. Some of the species also produce resin when the bark is cut, and in Cambodia one species in particular is very important to local communities as a source of the resin. The scientific name of this species is *Dipterocarpus alatus, known locally as Chhoeuteal* and in the timber trade by the name of *Keruing*. It usually grows into a big tree with beautiful dark red, hard and durable timber as well as producing prolific quantities of resin when injured. The resin is widely used in the region for caulking boats, much like tar, and so it provides an important source of supplementary income for villagers. The local people have developed techniques for harvesting the resin that does not kill the tree, and local custom recognises ownership of trees that are tapped for resin, by the person who first selected the tree.

After the Khmer Rouge were defeated by the Vietnamese there was a sort of peace, but remnants of the KR survived until about 1990 in the remoter forest areas in the east of the country in Ratanakiri and Mondulkiri Provinces. Here there were dense forests, rich in wildlife that had been home to ethic minorities. Many of these were displaced during the war, but began to move back when conditions became more peaceful. The one good thing about the presence of the KR was the general lack of disturbance of the forest, which had benefited the wildlife.

Figure 31: Undisturbed forest north of the town of Son Monorem in Mondulkiri Province

When real peace was finally restored to Cambodia, economic development became an obvious priority and the forests represented an easy source of quick money. Many foreign companies from Malaysia, China and Vietnam especially, were awarded concessions for logging, without any regard for the local people. There were also economic concessions that allowed forest to be cleared for establishing plantations of tree crops, including rubber, as well as other cash crops. Other 'developments' that expanded were mining and the construction of dams for hydro=power generation, all of which involved clearing some forest and by improving access also encouraged in-migration that led to further clearance of the forest. All this activity resulted in hardship for local communities as their forests were logged or cleared and they lost access to timber for house building and resin for cash, while the country gained little financial benefit as concessionaires found ways of circumventing rules and regulations and the government had neither the capacity or the will to do anything about it and most of the profits were transferred abroad.

The forests of Ratanakiri and Mondulkiri, adjoining Vietnam became a major target for both officially sanctioned and illegal logging. Stories abound of hundreds of trucks taking logs over the border to Vietnam to supply the rapidly expanding furniture manufacturing industry there. These logs included many resin trees that had been an important source of income, and while some were cut with the permission of the 'owner' in return for some cash, most were just cut regardless.

The widespread pillaging of the forest during the mid 1990s caused international outrage and pressure was brought to bear on the government to put a stop to it, which resulted initially in a moratorium on logging and eventually the cancelling of many concessions. The legal status of the forests in Mondulkiri and Ratanakiri has been clarified to some extent by declaring various areas as Biodiversity Conservation Areas, Forest Protected Areas and Wildlife Sanctuaries. The first two of these are managed by the Forest Administration while the third comes under the jurisdiction of the Ministry of Environment, but this does not seem to prevent economic concessions being awarded by the central government regardless of the official status of the forest. It is therefore common to find a rubber plantation in the middle of an area that is meant to be protected forest.

The forests in this part of Cambodia are particularly important for biodiversity conservation, since there are still a few remnants of good natural forest nearby in Vietnam and Lao PDR, some of which also have protected status as national parks or wildlife reserves. The Asian Development Bank is supporting a regional Project jointly between the governments of the three countries that aims to create a series of biodiversity corridors along both sides of the international borders. If it is successful in protecting the forests from further encroachment and can promote the restoration of the degraded forest lying between the protected areas it could help to restore the connectivity within this once vast natural forest and provide a secure habitat for a great number of species that are now endangered. This could be an extremely good use of the proposed REDD funds, and would bring many benefits to the local communities through employment and help with developing small enterprises based on the sustainable use of the local resources. Here's hoping it succeeds.

Figure 32: Squared baulks of timber from illegally felled trees destined for Vietnam and impounded by Cambodian Authorities (the tip of the iceberg)

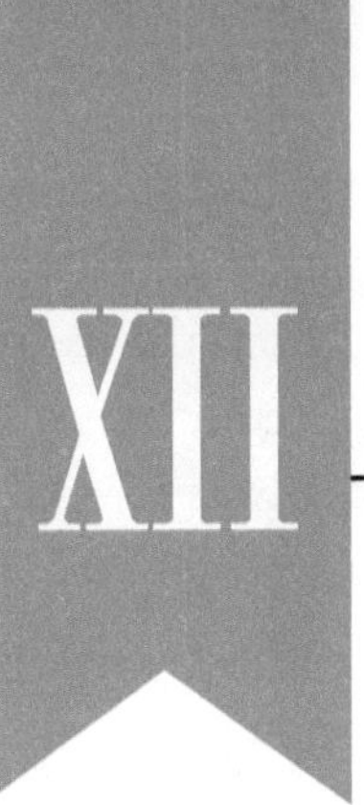

What next?

The Millennium Development Goals for reducing poverty and other laudable aims declared with such a fanfare by the UN in late 2000 are rarely mentioned these days, though there was a meeting in 2010 to assess progress, which seemed to suggest that some of the goals had been partially achieved, but more as a result of overall global economic growth rather than any specific 'development' initiatives. The UN Secretary-General at the time had to remind the world that with only five years left, it seems rather unlikely that the goals will be achieved and the global economic recession has recently reversed some of the earlier positive trends. The five years have now passed and little has been heard of the final outcome of the initiative.

Climate change pushed its way to the top of the political agenda, spurred on by Al Gore and his *Inconvenient Truth*, but already memories of hurricane Katrina that devastated New Orleans, two other severe typhoons that did as much damage in the Philippines in September 2009 and forest fires in many countries, that have been attributed to climate change, are fading. Climate change is gradually being replaced by the global financial crisis and recession in USA and Europe and terrorism as the top priorities. Most western countries' budgets are under severe strain as a result, and it seems likely that funds to support measures for dealing with the possible impacts of climate change will be cut back. Perhaps the lack of success at the Copenhagen Conference and all the controversy surrounding the veracity of the data was an indicator of the lack of conviction that it is really necessary for mankind to change lifestyle and adapt to the different conditions.

It seems strange that so much effort and finance is put into reducing the twenty per cent of greenhouse gas emissions that come from forests, and so little goes into the sixty per cent or so that comes from the use of fossil fuels. It is ironic that the melting of the ice in the Arctic ocean is seen as a golden opportunity to drill for even more oil that was formerly inaccessible below the ice. Presumably the oil companies want to melt as much of the ice as possible, regardless of its potential negative impact on climate in much of the rest of the world

Opinion polls in Europe and the USA are indicating an increasing level of scepticism regarding climate change among the general populace, and so, even if politicians are truly committed to doing something about it, they will have increasing difficulty in convincing their electorate, and so we can expect to see a decline in spending on efforts to adapt to, or mitigate the effects of, climate change, at least until there is another disaster somewhere that seems to be climate related.

The Future for Forestry

China accounts for about one sixth of the world's population, and with it's economy growing at around seven per cent annually, it needs huge and growing quantities of raw materials. If we just look at the use of wood for utility purposes and exclude fuel for energy, we find that the 1.2 billion people of China currently have an average consumption of about 0.19 cubic metres of logs per person per year, while a similar number of people in the whole of North America and Europe combined consume an average of about 0.53 cubic metres of logs per person per year, or about three times as much. This latter figure includes several poor countries as well as the rich ones, and if we look just at the USA and the EU we find that they consume about 1.57 and 0.74 cubic metres per person annually respectively while Japan uses about 1.3 cubic metres per person per year. These figures take account of trade so that they include imports and deduct exports. The point of all this is that if China achieves its aim of matching the 'west' in terms of living standards, its consumption of wood could increase five to tenfold from the current level. The question is: where will all this wood come from? China would need almost as much wood as the whole world consumes at the moment, over 1.5 billion m³ annually, and that assumes that population will not grow much more. Then what about India, with a similar population?

The world's forests, especially in the tropics, are under threat from many directions. Forests are cleared to release land for agriculture and growing other commodities such as rubber and oil palm as well as from logging, both legal and illegal, to meet the growing demand for wood for everything from house-building to furniture and paper. The flurry of excitement about climate change has turned attention to the fact that trees and the soils in which they grow, especially where they are peat, contain large amounts of carbon, which converts to carbon dioxide (a greenhouse gas) when burnt or allowed to decay which is why almost one fifth of the current emissions of carbon dioxide come from the clearance or degradation of forests,

So we are back to where we were in 1980 talking about the loss of the world's forests, but this time we are talking about carbon, instead of the forests *per* se and habitat destruction for all the biodiversity that inhabits them. The climate change economists say that reducing the rate of forest loss and degradation is the least cost way of reducing emissions of carbon dioxide, but it seems that their calculations have not really investigated the real cost. It is not just a matter of paying people a little money to protect an area of forest, as account has to be taken of the factors that are driving the forest destruction. If an area of forest is cleared and converted to oil palm or some other commodity, or as in parts of the Philippines, vegetables, the profits that can be generated for the grower are far larger than anything that the international community is willing to pay to keep the forest. The profit from converting forest to some other use is captured mainly by the big investors, while the local communities, at best, get a little employment. Even in the Philippines many of the farmers that clear the forest to grow vegetables are in *hoc* to middlemen who supply them with the seed and fertiliser and then undertake all the marketing.

So long as the world demand for commodities like wood, palm oil, rubber and so on continues to grow, forests will be under threat but the world also needs a wide range of goods and services that are provided by forests. Although the link between loss of forest cover and damaging floods is controversial, recent work covering fifty six tropical countries

(Corey *et al* 2007; see bibliography) has shown that there is a link, such that a ten per cent loss of forest cover can result in up to a twenty eight per cent increase in flood frequency. Natural forest, with its species diversity and irregular structure is found to be substantially better than plantations. Apart from the impact of reduction in forest area on floods, there is clear evidence that, on steep land, when forest is converted to agriculture, especially annual crops, the severity of erosion increases dramatically. This not only reduces soil fertility and hence productivity, but it also exacerbates flooding by silting up rivers and reservoirs downstream and promoting flash floods. Many important hydropower dams in the Philippines and other countries have been seriously impaired as a result of sedimentation of their reservoirs, and this represents a huge cost to the economy. The silt brought down by erosion, not only fills up the reservoir, but also increases wear and tear on the turbine blades.

Figure 33: Denuded catchment of Binga hydro-electricity reservoir with serious erosion in Luzon, Philippines

The world's demand for wood and wood products could ideally be met mainly from plantation grown wood, with just some of the wood with special properties, such as those required for musical instruments or durability and decorative uses coming from sustainably managed natural forests, which could be managed for multiple purposes, including the conservation of habitats and biodiversity. There is plenty of evidence to show that productivity of plantation forests can be increased markedly through a combination of careful selection of the genetic quality of the species grown and attention to protection and maintenance and crop nutrition. According to the FAO statistics there are about one hundred million hectares of plantations for timber production around the world, but more than half the area is in three countries, USA, China and Russia and much of the rest being of dubious quality and low productivity. In many countries the areas are too small to supply any sort of processing industry, other than small sawmills, so that they are hardly commercially viable. Comparison of three key indicators; population, log consumption and the area of industrial plantations, between the relatively developed and wealthy grouping of North America and Europe, and the less developed and poorer region of Asia-Pacific is interesting. Both regions account for just over one-third of the industrial plantation area, but the former has just twenty per cent of the population and consumes sixty seven per cent of industrial logs, while the figures are almost reversed for the latter with fifty six per cent of the population and eighteen per cent of the log consumption. This suggests that Europe and North America will probably be able to meet most of its wood needs from its own resources, while Asia-Pacific will have to establish huge areas of plantations to meet its needs.

While much attention is focussed on the forests and what is happening to them, little attention is paid to how the wood is used. When people buy a product made of wood, they may ask if the wood has come from a sustainably managed forest, but rarely, if ever, ask how much wood has gone into its production. Logs are mainly processed in the countries where they grow with less that ten per cent of logs being traded. About forty per cent of the logs produced are then either sawn or made into a variety of panels like plywood and particleboard. The rest is pulped and used for making paper. The sawn-wood is either used directly for construction or is further processed into doors and windows and furniture and thousands of other smaller products. Processors in developing countries are generally very inefficient and so add less value to the logs than if the logs are processed in a developed industrialised country.

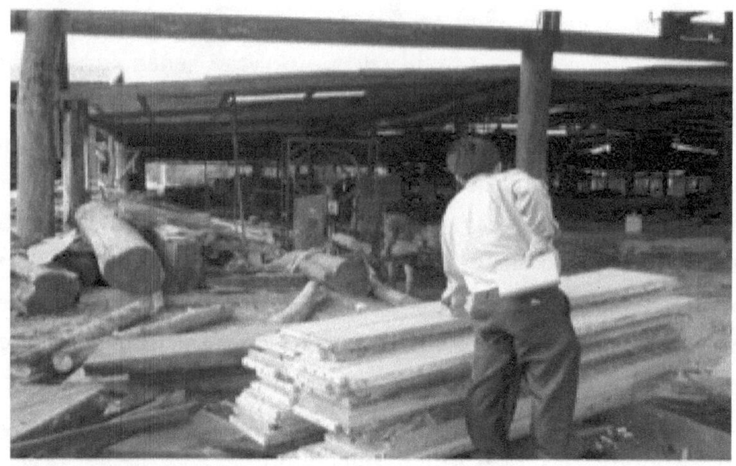

Figure 34. Typical small sawmill in a developing country with high proportion of waste and no attention to health and safety.

Based on general statistics from FAO on reported production of logs and different products for about two hundred countries it can be seen that on average in North America, only about 1.2 cubic metres of logs are required to produce 1 cubic metre of sawn-wood, while in Europe it is 1.45 cubic metres, in Asia-Pacific 2.12 cubic metres, in South America it is 2.46 cubic metres and in Africa it is 5.43 cubic metres. There is clearly plenty of scope for reducing considerably the volume of logs needed to produce the quantity of products that are needed.

The long established principle of forest management that will achieve sustainability, works on a rotation, so that only about three to five per cent of the forest is harvested each year, and the harvesting operations progress round the forest, returning to the same area every twenty to thirty years. This allows the forest to recover and replace the volume that was logged through growth. In most tropical countries a form of 'selective logging' is used, which means that only a few large trees that may only make up about five per cent of the total number of trees in the area are cut down. Some of the remaining trees may be damaged, but young saplings that have not had enough light to grow are released and rapidly fill up the gaps made where trees have been felled. This more or less simulates the natural process, where large old trees die and are eventually blown over, leaving gaps for regeneration to grow. Problems only arise if people take advantage of the easier access that the logging roads create, and move in to clear the forest for agriculture. This is all too common throughout the tropics where rapid population growth has driven the need for more land. In principle, if selective logging is practised and encroachment is avoided there is evidence to show that the harvesting can continue indefinitely so long as the total area of the forest is unchanged. This is where the system breaks down, because although the principles may be applied in theory, in practice the forest is subjected to encroachment and illegal logging occurs and the forest does not get a chance to recover, while areas due for re-logging may no longer be forest when the manager returns for a second harvest. Logging or harvesting a forest does not necessarily mean that the forest will be degraded or destroyed, as demonstrated in many European countries where the forests are not subject to illegal logging and encroachment.

In many European countries the area of forests declined steadily until about the middle of the nineteenth century or the early part of the twentieth century. Shortly after the second World War, Anderson (1950) studied the impact of state control of private forestry in eight European countries where a high proportion of the forests were in private ownership. The study looked at the evolution of forest policy and state attempts to control the activities of private forest owners, over the period from the mid nineteenth to the mid twentieth century, during which forestry became increasingly politicised and regulated. The study showed that from about 1850 onwards, increasing pressure on each country's forest resources led to legislation to restrict the rights of private owners. The cause of pressure on the resources differed between countries, including the growth of a pulp industry (Sweden), growth in sawmilling to supply timber for construction (Finland), growth in exports to feed growing demand in other countries (Norway), war (Belgium and Netherlands) and forest clearance to expand agriculture (France and Switzerland). An average of four Legislative Acts per country were introduced, with an average interval of eleven years between them. The main emphasis initially in most countries was to define protection forest for soil and water conservation, but gradually restrictions were also put on felling and harvesting, and requirements to regenerate forests were imposed.

The aim of all the policies was to prevent degradation and destruction of the forests and encourage regeneration in order to have continued supplies of raw material for the future. Most countries provided some form of subsidy to private forest owners towards the cost of management and/or regeneration and some also for afforestation of wasteland. It is interesting to note the evolution of policy follows a very similar course in all countries, and only differs in the timing. The first stage is restriction of grazing rights and common use rights in forests, and the designation of specific 'protection' forests. This was followed by restrictions on felling and forest clearance, which were gradually strengthened over the years. Then came moves to promote joint management of small privately owned forests and the rise in Associations or Cooperatives to provide the expertise and improve marketing. Finally measures were introduced to support afforestation on wasteland.

The accumulated measures had reached the point of a fairly comprehensive forest policy in all countries by 1946. Comparison of a number of indicators of forestry sector performance between then and 1980, when the effects of the policy should have had time to impact, shows significant changes. The main changes were broadly similar in all countries, and resulted in an expansion in the total area of forest by twenty three per cent over the thirty five years (1945-80) with privately owned forests accounting for more than eighty per cent of the increase. While population density increased by almost thirty per cent, forest area per capita only declined by four per cent. Consumption of industrial wood increased by only six per cent though population increased by thirty per cent.. In 1980 the area of forest classed as less than thirty five years old was eight per cent more than the additional forest area, confirming that the expansion was real, and that only modest areas of older forest had been felled and regenerated. The sustainable management of a forest requires that the area under management is fixed in perpetuity, since the volume of logs that can be safely removed each year must be balanced with the total annual growth throughout the rest of the forest.

While Europe has largely solved the problem of conserving its forests and ensuring sustainable supplies of wood and environmental services for the future, the same cannot be said for developing countries, which have to cope with the demands of population and economic growth.

What About 'Food Security'.

So, if climate change is the current *flavour of the month* what will the next one be and when will it emerge? There seems a good likelihood that it will be food security. In 2008 there was a rapid increase in food prices around the world covering onions in Bombay and rice in China, among other things, and FAO was prompted to issue warnings, and the G8 meeting in 2009 put food security at the top of the agenda along with climate change. The global economic recession put, what turned out to be, a temporary damper on the rise in prices, but in 2010, prices started to rise again, above the 2008 peak to a new peak in early 2011. In February 2011, the *Economist* ran a special feature on food security and pointed out that by the end of this year, the world's population will have reached seven billion and can be expected to reach nine billion by 2050, only forty years away. The recent rise in food prices has plunged many people back into poverty, undoing much of the apparent achievements attributed to the Millennium Goals.

The *Economist* considered that the main issues contributing to the increase in food prices apart from growth in demand due to population increase, are a declining rate of improvement in crop productivity, partly resulting from the increasing cost of chemical inputs of fertilisers and pesticides, climate related problems such as drought that have prompted export bans, changes in eating patterns because as developing countries become richer they increase the proportion of animal products, especially meat in their diets and the growth in the production of crops for non-food use, such as energy and fibre.

Regardless of these external factors that affect productivity, the increase in population alone will mean that production of food will have to increase by about thirty per cent in the next forty years just to feed the additional mouths at current levels. In the past, increases in output of agricultural products has been achieved by a combination of expansion of cultivated land and increased productivity, and data from FAO indicate that between 1980 and 2005 the area used for the cultivation of the twenty most important crops increased by about three hundred million hectares. During the same period their data, derived using different methods, showed that the area of forest had declined by about three hundred million hectares. The fact that the two figures are so similar lends some credibility to them. This three hundred million hectares represents a six per cent increase in the area under cultivation and although this is relatively small compared with the fifty five per cent increase in output over the same period, most of the increase in both output and area under cultivation occurred during the first decade and the rate of change in both has subsequently slowed substantially.

It seems that in the first five years of the millennium the rate of increase in output was lower than the rate of population growth. If forests are to be conserved then agriculture must rely on increasing yields on existing land rather than on further expansion of cultivated land. In fact the area for cultivating food crops may have to decline, since some of the forest land previously cleared will need to be replanted as forest for environmental reasons as well as some that will need to be planted for timber production to sustain supplies of wood while conserving the remaining natural forest. Thus it will no longer be rational to look at forest and agricultural strategies separately as is currently done, but it will be essential that land-use planning and investment strategies for both sectors are closely integrated in order to ensure the best possible use of land resources. The concept that forestry should account for a specific percentage of a country's land area,and the idea that forests should be the residual land after all other needs are met are not tenable and should be dropped.

Finally, a high (but unknown) proportion of the remaining natural forests are in hilly or mountainous areas with moderate to steep slopes, and the land is therefore generally unsuited to agriculture, and retaining forest cover is important for soil and water conservation. There are increasing worries in many countries that water supply will become increasingly difficult. Although forests transpire rather more water than grassland, they increase infiltration and because soils under forests are generally drier to greater depth than soils under arable crops they reduce the risk of flash floods and so improve the distribution of water over time. Water is needed in many areas for irrigation and this will be an important element of achieving food security.

Recent work described earlier to prepare countries for REDD+ has shown that policies for forestry, agriculture, water, energy and other sectors that use land, are very sector driven with little if any interaction between the sectors in drawing up strategies for development. In fact, despite some sector development plans and strategies, much development is very *ad hoc* and is driven by private interests of both small farmers and big business. Corruption often plays a big part in decisions to allow land to be converted from one use to another; usually from forest to something else, without regard for the social or environmental consequences. The strategic documents prepared to access REDD+ funds generally refer to the need to improve land use planning and inter-sectoral coordination, but only time will tell if these commitments can be put into practice.

Conclusions

Foresters generally are far too insular and tend to be 'behind the loop' in current jargon, always trying to catch up with what is going on, rather than helping to determine the agenda for change and development. However, there is an alarming trend becoming apparent whereby 'forestry' is being downgraded politically in an increasing number of countries. This is manifest by the separation of responsibility for conservation and environmental issues from the more commercial issues such as timber production and plantation establishment. It is difficult to know if this is deliberate on behalf of politicians to reduce the status and weaken forestry interests, or is merely a misguided attempt to put more emphasis on environmental issues. But in most cases forestry becomes a subsidiary part of a broad environmental mandate and gets lost among pollution and water quality concerns.

Foresters are partly to blame for this as most see themselves as servants of the politicians rather than makers of policy for the politicians to enact. In the UK it was public protest rather than professional advice that forced the government to rethink its policy on privatising state forests. In most developing countries, civil society is not strong enough to make policy and usually takes the role of opposing government proposals, which sets it on the opposite side from the forest authority in advocating what is required to implement policy, which is often ill-informed. In many countries, government policy is a 'knee-jerk' reaction to an event such as a flood with the introduction of a logging ban, while in others it is based on spurious and unjustified targets for having a stated per cent of the land area as forest, often without saying how forest is defined.

In 2001, I was involved in crafting a forest policy for the Asian Development Bank as a basis for guiding its investments in forestry. The outcome was a disappointment, because it was never adopted by the Board, on the grounds that there were too many sector policies, and investment in large scale infrastructure projects is more appropriate for the Bank.

As I hope this book has shown, forestry interests touch on all aspects of human welfare in one way or another, and foresters need to be at the forefront in determining the next *flavour of the month*. Management in forestry needs to be about managing change and new ways need to be found of achieving sustainability (not stability) in a changing world. Many young people going into forestry have tended to specialise in some aspect, such as community forestry or conservation and while we need people with in depth knowledge and experience in certain topics, anyone with ambitions to lead the sector must develop a broad understanding not just of the trees and the wildlife but also how forests and the products and services that forests provide interact with all aspects of human society. I hope that many foresters in the future will have as interesting a career as I have had

Bibliography

Anderson, M. L., (1950), *A survey of development in eight countries of western Europe in the decade 1938-1948*, Oxford University Press, Oxford

Becker, E. (1986), *When the war was over: Cambodia and the Khmer Rouge*, Simon and Schuster, New York. ISBN 0-671-41787-8

Cibula, E. (1980) *Trends in timber supply and trade: an information review*, BRE UK Dept of Environment.

Cooke W., and U. Kothari. (2001) *Participation: the new tyranny*, Zed Books:London

Corey J., A. Bradshaw, N. S. Sodhi, K. S.H. Peh and B. W. Brook (2007). *Global evidence that deforestation amplifies flood risk and severity in the developing world*. Global Change Biology (2007) 13, 1—17.

Deutsch, W. P., A. Busby, J.L. Oprecio, J.P. Bago-Labis and E.Y. Cequina, (1998) Community based water-quality monitoring from data collection to sustainable management of water resources. Chapter 7 in Seeking Sustainability.

Ernsting, A. and D. Rughani, (2007) *Reduced Emissions From Deforestation': Can Carbon Trading Save Our Ecosystems?*, www.biofuelwatch.org.uk/category/reports/page22/

Fraser, A. I. (2002),*Making Forest Policy Work*, Kluwer Academic Publishers, Dordrecht. ISBN 1-4020-1088-5.

Gammie, J.I. (1981), *World timber to the year 2000*, Economist Intelligence Unit.

Kenney-Lazar, M. (2010) *Land concessions, Land tenure, and livelihood change: plantation development in Attapeu Province, Southern Laos*, Fulbright Scholarship Report.

Meyfroidt, P and E. F. Lambin (2009), *Forest transition in Vietnamand displacement of deforestation abroad.* Arizona State University, www.pnas.org/doi/10.1073/pnas.0904942106

Stern, N.,(2006) *Review on the economics of climate change*, final report, http://www.hmtreasury.gov.uk/independent_reviews/stern_review_economics_climate_change/stern_review_report.cfm

Index